4th International Scientific and Technical Conference on Scientific and Technical Progress in Ferrous Metallurgy (SATPIFM-2019)

IOP Conference Series: Materials Science and Engineering
Volume 718

Cherepovets, Russia
18 - 20 September 2019

Editor:

Alexsandr Kozhevnikov

ISBN: 978-1-7138-2577-7
ISSN: 1757-8981

Printed from e-media with permission by:

Curran Associates, Inc.
57 Morehouse Lane
Red Hook, NY 12571

Some format issues inherent in the e-media version may also appear in this print version.

Institute of Physics
Dirac House, Temple Back
Bristol BS1 6BE UK

Phone: 44 1 17 929 7481
Fax: 44 1 17 920 0979

techtracking@iop.org

Additional copies of this publication are available from:

Curran Associates, Inc.
57 Morehouse Lane
Red Hook, NY 12571 USA
Phone: 845-758-0400
Fax: 845-758-2633
Email: curran@proceedings.com
Web: www.proceedings.com

TABLE OF CONTENTS

MATERIAL SCIENCE

EDUCATION

IOP Conf. Series: Materials Science and Engineering **718** (2020) 011001 doi:10.1088/1757-899X/718/1/011001

4th International Scientific and Technical Conference on Scientific and Technical Progress in Ferrous Metallurgy (SATPIFM-2019)
18-20 September 2019, Cherepovets, Russia

Preface

The 4th International Scientific and Technical Conference on Scientific and Technical Progress in Ferrous Metallurgy (SATPIFM-2019) was held in Cherepovets, Vologda Region, Russia on September 18th - 20th, 2019. The organizers of the Conference are: PAO Severstal and Federal State Budgetary Educational Institution of Higher Education "Cherepovets State University".

SATPIFM is the regular international conference series in Russia. Initially it was held in Cherepovets biannually since 2013.

The goal of the conference is to bring together scientific and research teams in the field of ferrous metallurgy, as well as the representatives of the enterprises that operate in the steelmaking industry, in order to exchange innovations and to implement them, to search for the best proposals and ideas for solving urgent global problems for the metallurgical industry, to develop a scientific and research network, collaborations and international cooperation.

The conference was attended by more than 150 representatives of 12 countries (Russia, Ukraine, Belarus, Kazakhstan, Latvia, France, Belgium, Germany, Poland, Israel, Switzerland, Japan), 17 universities (Cherepovets University, Vologda University, Higher School of Economics, National University of Science and Technology MISiS, Bauman Moscow State Technical University, Moscow Polytechnic University, Moscow Aviation Institute, Peoples' Friendship University of Russia, Skolkovo Institute of Science and Technology, Lipetsk University, Nosov Magnitogorsk State Technical University, Ural Federal University, Nizhny Tagil Institute of Technology, Donetsk National Technical University, Dahl Luhansk National University, Rudny Industrial Institute, Karaganda State Industrial University), 14 manufacturing enterprises (Cherepovets Steel Mill, Magnitogorsk Iron and Steel Works, Novolipetsk Steel Plant, EVRAZ, Nippon Steel, Accusteel Ltd., Chepetsky Mechanical Plant, Volzhsky Pipe Plant, Ashinsky Metallurgical Works, Chelyabinsk Pipe Rolling Plant, Pervouralsk New Pipe Plant, KAMAZ, Illich Mariupol Steel and Iron Works, Yenakiyeve Iron and Steel Works) and more than 40 scientific and research, design, research and production organizations (I. P. Bardin Central Research Institute of Ferrous Metallurgy, National Research Centre "Kurchatov Institute", Prometey Central Scientific Research Institute, Institute of Strength Physics and Materials Science of the Siberian Division of the Russian Academy of Sciences, Institute of Metallurgy of the Ural Division of the Russian Academy of Sciences, SMS Digital GmbH, SIEMENS, Schneider Electric, Steelmasters bvba, Sojitz LLC, Research Network Lukasiewicz - Tele & Radio Research Institute, Severstal-Project, Research Centre "Metallurgy of High Technologies", Ferrous Metallurgy Institute of the National Academy of Sciences of Ukraine, Lechler GmbH, FIVES KEODS, Severstal-Infocom, ZAO Prochnost, etc.).

This collection of articles includes the articles which were selected by the experts of the organizing committee from a large number of submitted reports and speeches.

The conference organizing committee would like to thank all the guests of the conference for their participation, prepared reports and the publication of the latest achievements of science and technology in the metallurgical field. I would like to thank all the guests and participants of the conference for their efficient and productive work, as well as the organizers for a careful planning of SATPIFM-2019, the experts and reviewers who made their valuable comments and suggestions.

Editor
Ph.D. Alexsandr Kozhevnikov
Cherepovets State University, Russia

IOP Conf. Series: Materials Science and Engineering **718** (2020) 011002 doi:10.1088/1757-899X/718/1/011002

Peer review statement

All papers published in this volume of *IOP Conference Series: Materials Science and Engineering* have been peer reviewed through processes administered by the proceedings Editors. Reviews were conducted by expert referees to the professional and scientific standards expected of a proceedings journal published by IOP Publishing.

IOP Conf. Series: Materials Science and Engineering **718** (2020) 012001 doi:10.1088/1757-899X/718/1/012001

Characteristics of High-Strength Sheet Steel after Subsequent Heating

V I Alimov[1], A P Shtykhno[1] and Yu S Kolchina[1]

[1] Donetsk National Technical University, 58 Artem Street, Donetsk, 283001, Ukraine

E-mail: alim41@mail.ru

Abstract. The influence of controlled rolling and the parameters of subsequent technological heating on the processes of structure formation in high-strength sheet steels EH36 and X70 is investigated. It has been established that when redistributing a thick sheet into products, it is necessary to minimize the temperature-time parameters to prevent softening processes. Welding, as one of the sources of technological heating, leads to a change in the structure in the zone of thermal influence and uneven distribution of hardness over the cross section of welded joints, which reduces the uniform strength of the structure as a whole. Short-term heating (about 15 minutes) even to temperatures above Ac1 (760 ° C) and cooling in air in the weld zone do not lead to significant changes in structure and hardness, but the degree of uniformity of the welded sheets increases.

The quality of sheet steel is determined by many factors – chemical and physical homogeneity, minimum content of contaminants and non-metallic inclusions, high surfacequality and the density of metal in general. It has been observed on numerous occasions that after controlled rolling there is an inhomogeneous structure in particular areas of sheets, predominantly with the size of ferritic grain 8-10 μm across, however, there are bigger ferritic grains up to 30 μm across. Besides that, over the thickness of the sheets there is often inhomogeneous crystallographic texture and especially strong ferrite-pearlite banding. Inhomogeneity of microstructure and of crystallographic texture in the state after controlled rolling inherently leads to anisotropy, decrease in impact strength and brittle fracture resistance of the metal [1-6].

The purpose of this study was to research the peculiarities of structure formation in high-strength sheet steels EH36 and X70 after controlled rolling and after subsequent technological heating.

The high-strength steels EH36 and X70 were researched, which are fine-grained ferrite-pearlite steels. The chemical composition of the steels under study is specified in Table 1.

Table 1. Chemical composition of steels under study, % wt.

Steel grade	C	Si	Mn	S	P	Cr+Ni+Cu	N	V	Ti	Nb	Mo	$C_{equivalent}$
EH36	0,11	0,21	1,18	<0,005	<0,014	0,08	0,003	0,25	0,015	0,04	0,05	<0,37
X70	0,10	0,35	1,20	<0,004	<0,012	0,47	0,008	0,050	0,010	0,04	0,1	<0,41

The microstructure of steels EH36 and X70 after controlled rolling (Figure 1 (a) and (b) respectively) is predominantly aferrite-pearlite mixture with strong pearlite banding, but there are also ferrite grains elongated along the rolling direction, in which the recrystallization processes did not finish in time. The main reason of banded structure formation with elongated ferrite and pearlite bands is supposed to be the inherent impact of deformed uncrystallized or partially recrystallized austenite on its decomposition during overcooling below critical points.

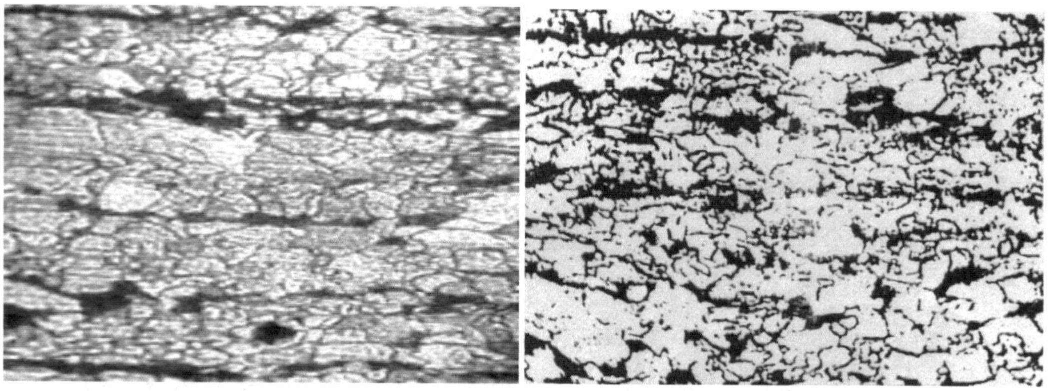

a b

Figure 1. Microstructure of steels EH36 (a) and X70 (b) after controlled rolling, x400.

It is known that the higher the overall degree of deformation of austenite at temperatures at which its recrystallization is hindered, or substantially restrained and the higher the elongation of its grains in the direction of metal deformation process, the longer the elongated bands of pearlite sectors will be after pearlitic transformation is finished.

The presence of pearlite banding leads to high anisotropy of mechanical properties. Along the normal to sheet surface plane the tensile strength, yield strength and percentage of elongation are almost twice lower than in directions in the plane of the sheet [5].

To study the influence of subsequent technological heating for heat treatment on the structure and properties of sheet steel EH36 the samples were heated up to 200 – 900 °C and held at those temperatures for up to 2 h and subsequently cooled by ambient air. The microstructures after additional heating are shown at Figure 2.

While heating up to 200 °C (Figure 2 (a)) the initial stage of the ageing process is happening, the microstructure and properties basically do not change, but in the process the carbides coherently bound with the matrix are precipitated, pearlite banding is maintained, and the hardness increases to 124 – 130 HB.

While heating up to 400 °C (Figure 2 (b)) substantial changes in microstructure are observed, which consist of rupture of pearlite bands, release of strain hardening after deformation; the level of residual stresses lowers and there is uniformly distributed equiaxial structure formed, which makes the hardness increase up to 136 – 145 HB.

At the temperature of 600 °C (Figure 2 (c)) there is partial recrystallization of the matrix, which is accompanied by the refinement of ferrite; during such heating the structure is formed in which fine particles of cementite (pearlite) are uniformly distributed in the ferrite matrix, and during that maximum hardness increase is observed (up to 150 – 155 HB), which is conditioned by the equiaxiality and dispersity of structure.

Heating and holding at the temperature of 750 °C (Figure 2 (d)) does not lead to substantial changes in the microstructure, but the initial partition processes of pearlite bands can be observed which, in turn, increases the hardness from 118 – 123 HB to 125 – 130 HB.

Heating up to 950 °C (Figure 2 (e)) leads to the manifestation of the influence of dendritic inhomogeneity and liquation, which substantially impacts mechanical properties, and the hardness lowers to 110 – 115 HB.

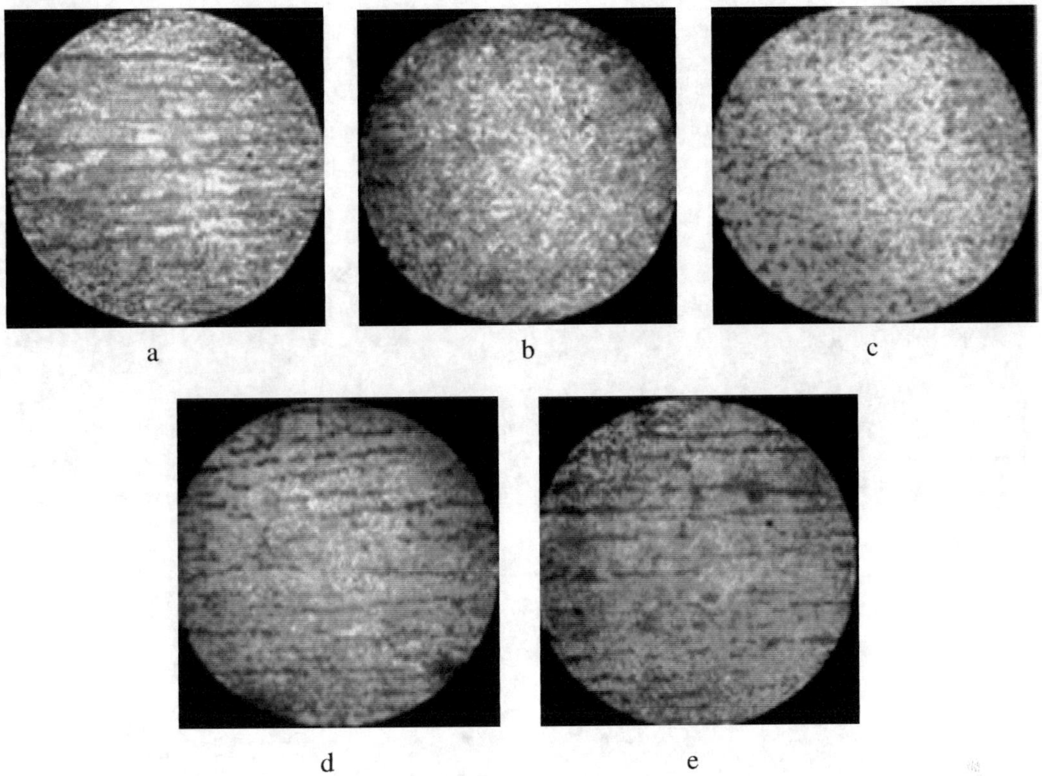

Figure 2. The microstructures of EH36 steel after additional heating up to the temperatures: (a) – 200 °C, 2 h; (b) – 400 °C, 1 h; (c) – 600 °C, 30 min; (d) – 750 °C, 20 min; (e) – 900 °C, 15 min; x 400.

The histogram Figure 3 shows that additional heating up to 200-900 °C with subsequent cooling by ambient air leads to non-monotonic change in hardness.

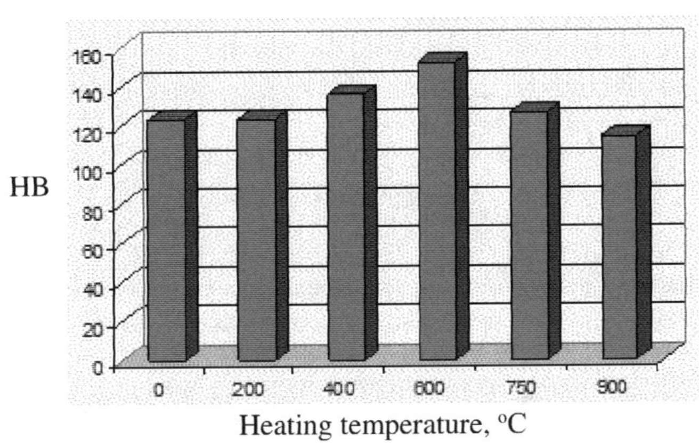

Figure 3. Hardness of sheet steel EH36 after additional heating.

Hardness increase while heating up to 600 °C can be conditioned by ageing and partial recrystallization, which is accompanied by the refinement of ferrite constituent.

The analysis of microstructural changes during additional heating showed the important role of structure formation during subsequent welding processes of different steel structures made of these steels.

The study of the influence of short-time heating during welding on the changes in structure and hardness over the weld section was done with the samples of sheet steels X70 and EH36, which underwent automatic hidden-arc welding and semi-automatic welding respectively.

The microstructure of a weld of sheet steel X70 is showed on Figure 4.

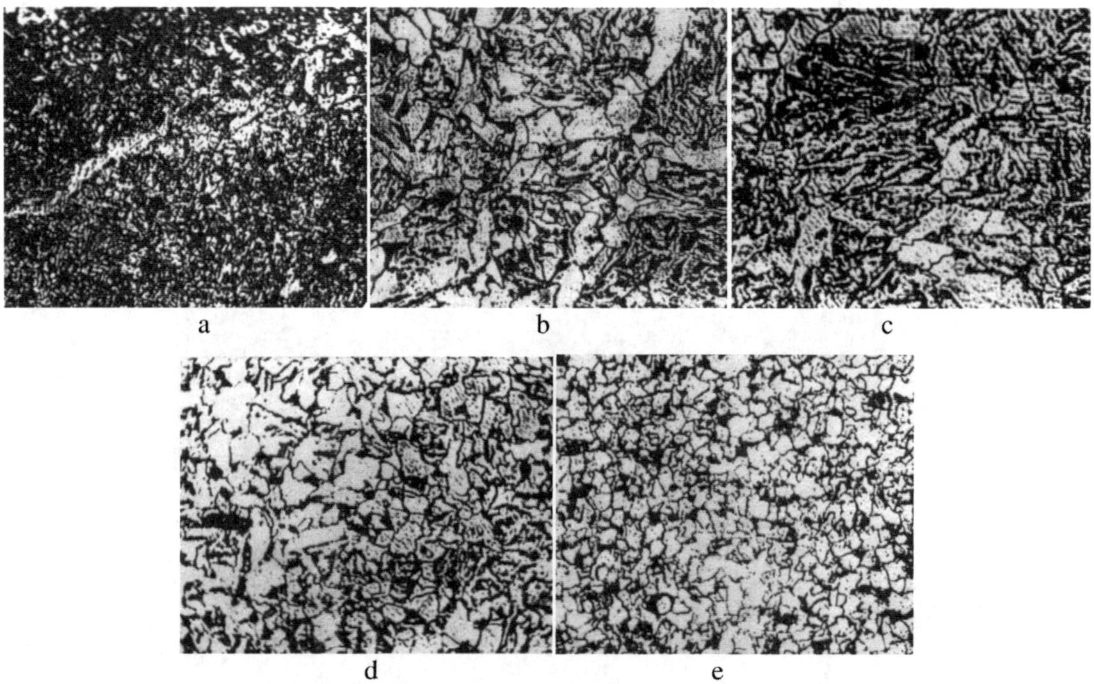

Figure 4. The microstructure of a weld of steel X70:
(a) – weld metal; (b) – weld area; (c), (d), (e) – heat-affected area, x500.

From the analysis of the microstructure of the studied samples we can see that depending on the influence of temperature gradient there are three zones clearly distinguishable in the welded connections: weld area, heat-affected area and the area of weld metal. Heat-affected area, where metal undergoes a particular thermal influence, deserves special attention. When moving away from the weld area to the main metal, the structure of metal changes according to the heating and cooling thermal cycles and depends on the chemical composition of metal and preliminary thermal and thermomechanical treatment.

The microstructure of weld metal and weld area (Figure 4 (a), (b)) consists of the products of intermediate breakdown of austenite and ferrite in the form of streaks on the borders of the crystals. In the weld area (fig.4b) the structure of martensite and bainite is also visible. Rapid growth of austenite grains happens in this area by means of heating up to a high temperature (1100 – 1150 °C).

In the heat – affected area we can see the change of microstructure from martensite-bainite to ferrite-pearlite, that corresponds to the processes of normalization, incomplete recrystallization and complete recrystallization during thermal treatment with special heating, that means that pearlite banding basically disappears in this area. Fine grains in the area of complete recrystallization are explained by the high speed of cooling down from the recrystallization temperature and the chemical composition of steel X70.

Figure 5 shows the dispersion of hardness by section of weld samples made of steels X70 and EH36.We can see on the graph that the highest hardness is observed in the central area of a weld seam. The reduction of hardness corresponds to the area of overheating and the sections of incomplete recrystallization.

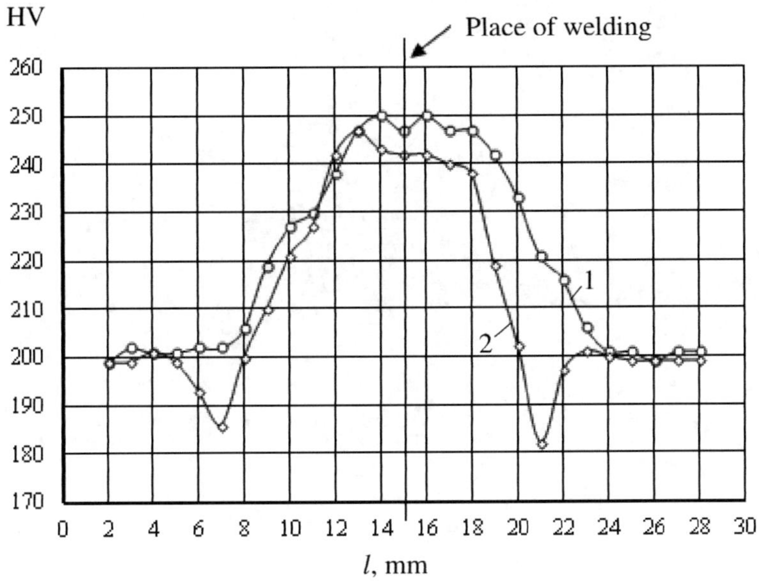

Figure 5. Dispersion of hardness HV over the weld connection section l of the samples: 1 – steel X70; 2 – steel EH36.

Such uneven dispersion of structure and properties over the weld connections section may cause concentration of high thermal stress and increase the risk of cracks and structural breaks in general.

Conclusion

Microstructure of steels EH36 and X70 in a state after controlled rolling is ferrite-pearlite with expressive banding of pearlite clusters and ferrite grains elongated along the rolling direction, where recrystallization processes did not finish in time.

The presence of pearlite banding can lead to anisotropy of mechanical properties. Along the normal to sheets surface planet he tensile strength, yield strength and percentage of elongation are almost twice lower than in directions in the plane of the sheet; the correction of this structural imperfection of steel EH36 after controlled rolling by an additional thermal treatment is possible.

Additional heating up to 200 – 900°C with subsequent cooling by ambient air leads to non-monotonic change of hardness.

Welding, as one of the types of technological heating, leads to the change in structure in the heat-affected area and uneven dispersion of hardness over the section of welded samples. The weld metal area has the highest hardness; such a change of properties in a weld seam may increase the risk of cracks and structural breaks in general.

References

[1] Pilguk V Ye, Alimov V I, Pilguk S V and Pasechnik S Yu 2018 *Low alloyed sheet steel* (Kharkov: FLP Panov A N) p 345

[2] Alimov V I and Georgiadu M V 2018 *Resourcesaving technologies of production and processing by pressure of materials in mechanical engineering* **1** 26–38

IOP Conf. Series: Materials Science and Engineering **718** (2020) 012001 doi:10.1088/1757-899X/718/1/012001

[3] Panasenko F L 1959 *Rolling and heat treatment of thick sheets* (Moscow: Metallurgizdat) p 154
[4] Brovman M Ya 1969 *Improved technology of rolling thick sheets* (Moscow: Metallurgy) p 387
[5] Stepanov A A 2001 *Metallurgy and heat treatment of metals* **7** 39–40
[6] Egiz I V and Shamray V F 2008 *Metallurgy and heat treatment of metals* **1** 33–36

Experimental Research of the Hot-Rolled Pipe Wall Sizing Process at the Reeling Mill of Pipe-Rolling Plant-140 in Conditions of Product Mix Extension

S Yu Belyaev[1], Yu B Chechulin[1], A A Fedulov[1] and M Yu Bulganina[1]

[1] Ural Federal University, 19 Mira street, Ekaterinburg, 620002, Russia

E-mail: belyaev_s_66@mail.ru

Abstract. This experimental study was conducted on the equipment of pipe-rolling plant-140 with an automatic mill in order to give a qualitative and quantitative evaluation of the parameters which define the pipe wall sizing process in reeling mills. During investigation values of the power parameters of the process were recorded for different sizing routes. Conclusions about the special features of the wall sizing process and feasibility of extension of the reeling mill process capabilities were formulated.

The main areas of development of modern pipe rolling plants are increasing pipe accuracy, mastering pipe production of high-strength steel grades and extending the range of pipe diameters. In the process routes of hot-rolled pipe manufacture the pipe wall thickness accuracy is obtained through wall sizing at reeling machines by screw rolling. At present, there are no comprehensive procedures in place to describe this process.

The subject of this research was the main lines of the work stand of the reeling mill which is part of the process line of pipe-rolling plant-140 with an automatic mill [1]. Initially, this pipe-rolling plant was intended for manufacture of pipes up to 140 mm in diameter. Later on, due to mastering of the production process after many years of operation the range of pipe diameters was extended up to 159 mm, with a wall thickness from 5 to 14 mm, for pipes of carbon an low-alloyed steel grades. Currently, production capabilities for manufacture of pipes up to 178 mm in diameter are being developed at pipe-rolling plant-140 without significant modification of the existing equipment. As a result, the equipment failure rate has increased dramatically due to the increased process loads. Such failures lead to long shutdowns of the entire pipe-rolling plant-140, which, for sure, impairs its technical and economic performance. To solve the task of ensuring stable manufacture of high-quality products at pipe-rolling plant-140 is possible only through collection and analysis of the fullest information on parameters characterizing operation of the electric drive, on mechanical transmissions and power parameters of the pipe-rolling process. This can be done only as part of a comprehensive experimental research.

Reeling mills of pipe-rolling plants with an automatic mill perform a process operation of sizing the wall of rough pipes rolled at the automatic mill. This process is carried out through screw rolling in rolls turned at a plug angle and feed angle relative to the pipe axis. During this process, a pipe workpiece is expanding in diameter on a tapered mandrel with a cylindrical belt and is being sized at the same time in radial clearances formed by the positional relationship between the rolls and the mandrel. After the reeling mill the pipe is transferred to the sizing mill to have its final diameter formed.

The reeling mill of pipe-rolling plant-140 is a non-reversing screw rolling mill with individual drives for each of its three rolls (Figure 1).

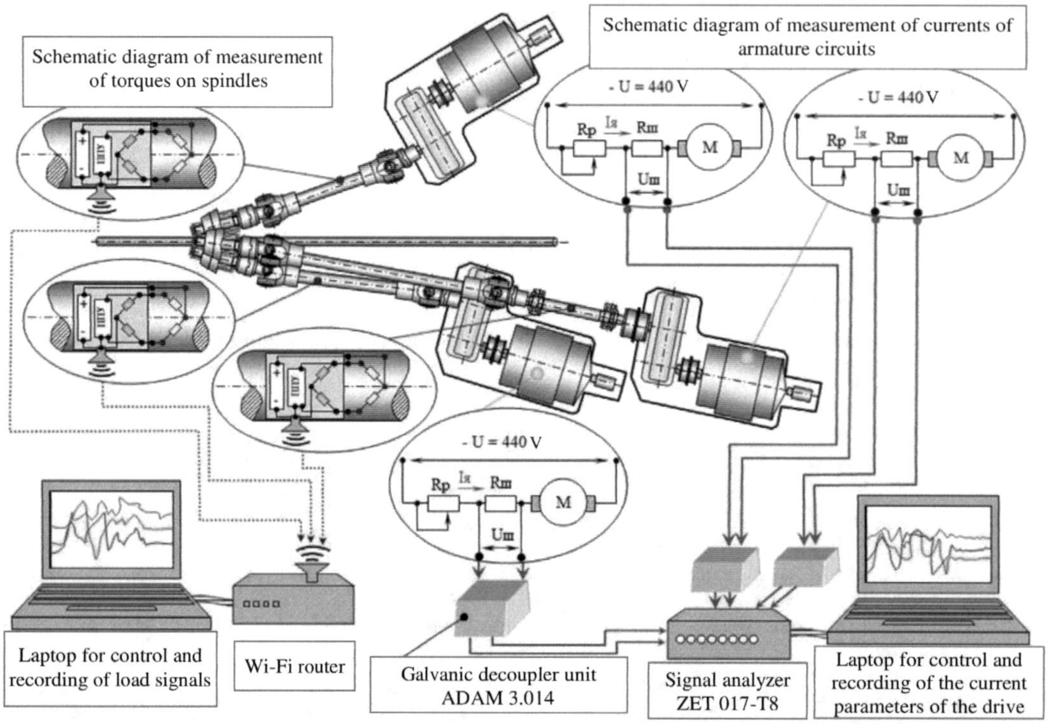

Figure 1. Drive design and schematic diagram of measurement and recording of load parameters in the drivelines of the work rolls of pipe-rolling plant-140 reeling mill during rolling.

Each of the rolls' drivelines includes a work roll assembly, universal spindles on rolling bearings, one-stage parallel-shaft reduction gear TsO-600, motor gear coupling M3-9, separately excited main drive motor P 143-4K with nominal power of 250 kW at nominal rotary speed of 500 rpm and constant voltage of 440 V. Drivelines of the bottom left and right rolls in the rolling direction are similar and differ only in the length of spindles and swivel twist angles. The driveline of the top roll includes additionally an intermediate shaft and connection gear coupling MZ-11. In line with the technical specification, the following nominal and maximum permissible load parameters for the reeling mill drive units are set:

- Nominal current/ cutoff current for the electric motor rotor – 612/1500 A,
- Nominal torque of the motor – 4.77 kN·m,
- Maximum torque transmitted by coupling MZ-9 – 29.4 kN·m,
- Maximum short-time/long-time torque transmitted by one spindle – 58.9/39.2 kN·m,
- Maximum short-time/long-time torque on the output shaft of the reduction gear – 31.4/16.7 kN·m,
- Maximum torque transmitted by coupling MZ-11 – 69.7 kN·m.

To identify the most loaded drive units of the reeling mill rolls, loading of each component was determined in reference to the threshold values when nominal torque of each of the main drives of the mill is applied. The value of the nominal torque equals 4.77 kN·m. The results of this analysis with regard to the reduction ratio are given below:

- Motor coupling MZ-9 – 16.2 %,
- Output shaft of the reduction gear by short-term load – 56.5 %,
- Output shaft of the reduction gear by long-term load –76%,
- Toothed coupling MZ-11 of the line's intermediate spindle's drive – 28%,
- Spindles by long-time torque – 45.2 %,
- Spindles by short-time torque – 30.1 %.

As it follows from the data given, the maximum load falls on the main electric drives.

The schematic diagram of measurement and recording of the load parameters in the drivelines of the reeling mill work rolls during rolling is given in Figure 1. Two methods of measurements were used simultaneously for recording of parameters:

– monitoring system for electric drives' current parameters,

– monitoring system with the use of torques – wireless data transfer system.

The system of measurement and recording of the current parameters of electric drives is built on the basis of 8-channel signal-analyzer ZET 017-T8, which operates in the mode of a multichannel oscillograph in combination with a laptop with input signal sampling frequency from 0.1 Hz to 100 kHz and an analog-to-digital converter with resolution by digital signal of 2^{16} (16-bit signal). All measured parameters are converted into digital form with the use of the analog-to-digital converter and amplifiers integrated in the analyzer, recorded by the computer and simultaneously visualized on the monitor screen.

The currents of the armature circuits of the electric drives of the reeling mill work rolls were recorded from armature shunts (75mV/1500A) of the armature windings of each of the electric drives. Signals were connected to signal analyzer ZET 017-T8 through galvanic decoupler unit ADAM 3.014, which ensured protection of the measuring circuit. To evaluate dynamics of the current loads of the electric drives of the drivelines of the reeling mill work rolls, the sampling rate for the recorded signal was 0.02-0.04 s, which corresponds to the frequency of 25-50 Hz.

The original system for measurement and recording of load parameters, which was developed in Ural Federal University, is a wireless data transfer system. The equipment set includes a transmitter-receiver with an independent power supply unit, a router and a laptop. The system is able to operate both with analog and digital physical measuring transducers. A PLC is used to collect data, which are subsequently transferred via Wi-Fi to the laptop, where the obtained data are recorded, visualized and analyzed.

Torques transmitted by spindles of the rolls' drive were measured by electrotensometry [2]. Sockets of resistance strain gauges (half-bridge circuit) of 2 gauges FK-20-100 and transmitter-receivers were installed on the smooth part of the spindles and intermediate shaft to record torques. Strain-gage sockets were powered from independent power supply units – accumulators – which were installed in the transmitter-receivers. Output signals from strain-gage sockets were sent to the PLCs of the transmitter-receivers, digitized and transmitted through a Wi-Fi network to the router and further to the laptop. To evaluate dynamics of the processes running in the drivelines of the sizing mill work stands, the sampling rate for the main power parameters was equal to 0.016 s (62.5 Hz).

In addition, for each batch or lot of pipe workpieces the following data were recorded in a table format: steel grade, rough pipe dimensions after the automatic mill, gauge size, pipe diameter and wall thickness after the reeling mill, mandrel diameter and setting, roll feed angle, rolling speed, special rolling conditions.

No special methods for planning of the experiment were used during inspection of pipe-rolling plant-140 reeling mill due to the intensive production. Instead, series of measurements were taken during regular production processes for different pipe sizes and steel grades. The following pipe sizing processes were recorded during the research:

– finished pipe diameters – 108, 140, 146, 159, 168, 178 mm,

– finished pipe wall thickness– 5.0 … 14.0 mm,

– pipe steel grades – 10, 20, 45, 09Г2С, 13KhFA, 30G2, 37G2F.

To decipher the obtained results, a software package which allows representing oscillograms in an easy-to-read format was developed. Analysis of the findings of the investigation included a qualitative analysis to identify the key features of pipe processing at the reeling mill, and a quantitative analysis to obtain the main relations between actual power parameters of the process and maximum permissible loads on the mechanical and electrical equipment.

Considering different steel grades, a wide range of diameters and different speed modes of electrical motors at the reeling mill, two most representative routes allowing evaluation of the influence of the mentioned factors were selected:

– pipe Ø 178 x 9.2 mm of steel grade 37G2F – this route is one of the most loaded because it is used for rolling of pipes of the largest diameter with a wall reduction of 0.8 mm of steel grade with the highest strain resistance (141– 211 MPa), and rolling is performed at speed corresponding to the first zone of electric motor speed control – not exceeding the nominal speed,

– pipe Ø 146 x 7.0 mm of steel grade 30G2 – this route is the most common route, it is carried out at increased rolling speeds corresponding to the second zone of electric motor speed control, with the strain resistance of steel within 128-185 MPa.

The strain resistance of steel was taken at the values of thermomechanical parameters [3] corresponding to the deformation conditions at the reeling mill: deformation degree of 10%, deformation rate of 10 s^{-1}, deformation temperature within the range 1000-900°C.

The qualitative analysis of the oscillograms for the batch of rolled pipes showed that there is a clear recurrence of the loading conditions between individual pipe workpieces, and that the torque values recorded with the use of the strain measurement method are consistent with the data obtained by recalculation through the currents of the electric motors' armature circuits.

The oscillograms of the electric motors' current loads, torques on the spindles in combination with the above mentioned limitations for different components of the reeling mill drive for the two most representative rolling routes are given in Figures 2 and 3.

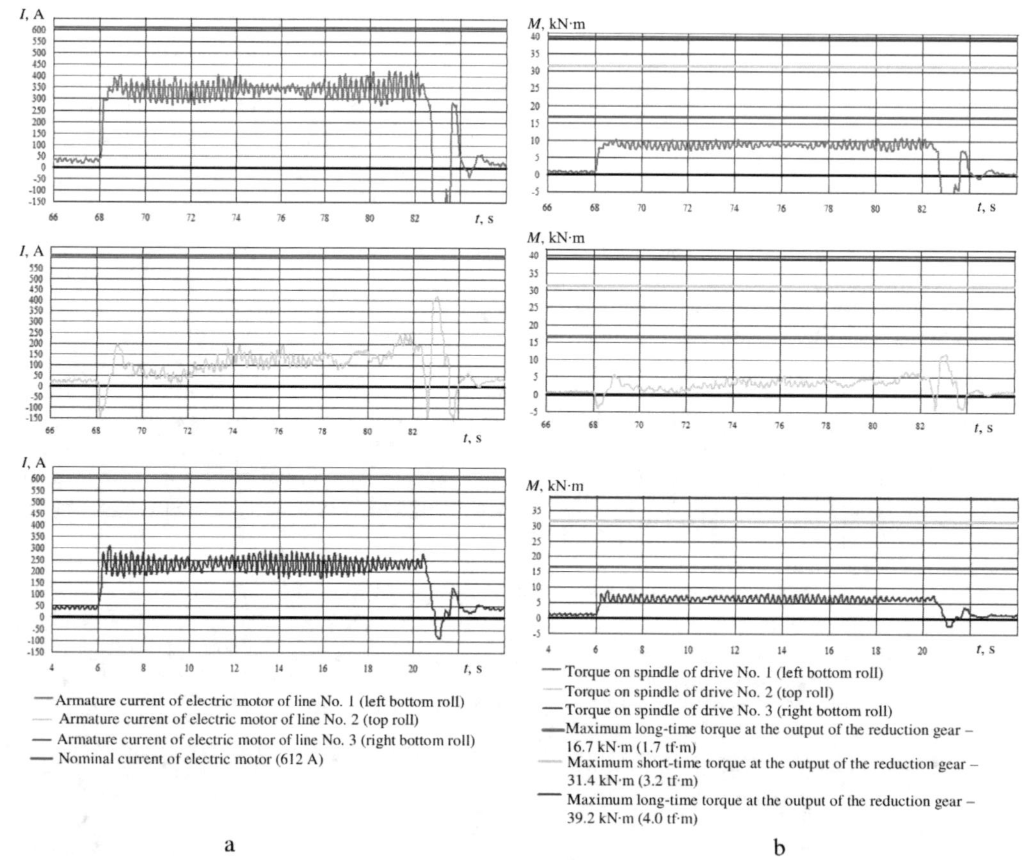

Figure 2. Oscillograms of the electric motors' currents *I* and torques on the reeling mill spindles *M* during sizing of pipe worpieces in the process of production of 178 mm diameter pipes with a wall thickness of 9.2 mm of steel grade 37G2F: (a) – oscillograms of the electric motors' drives; (b) – oscillograms of the torques on the spindles.

Analysis of the results of the investigation made it possible to come to the following conditions:

1. The pipe wall sizing process by the method of reeling on a retained mandrel in screw-rolling mills is a steady process and is characterized by significant oscillations of the torque and current values relative to the mean value. Oscillation amplitude amounts to 10 – 30 % of the mean load value, and oscillation frequency is double the rotation speed of a pipe workpiece – 4.0 … 6.0 Hz.

The cause of this oscillating process is deformation of the side angles of rough pipes rolled at the automatic mill. As a matter of fact, that is the main technological function of the reeling mill under investigation, and such loading conditions are standard for this mill.

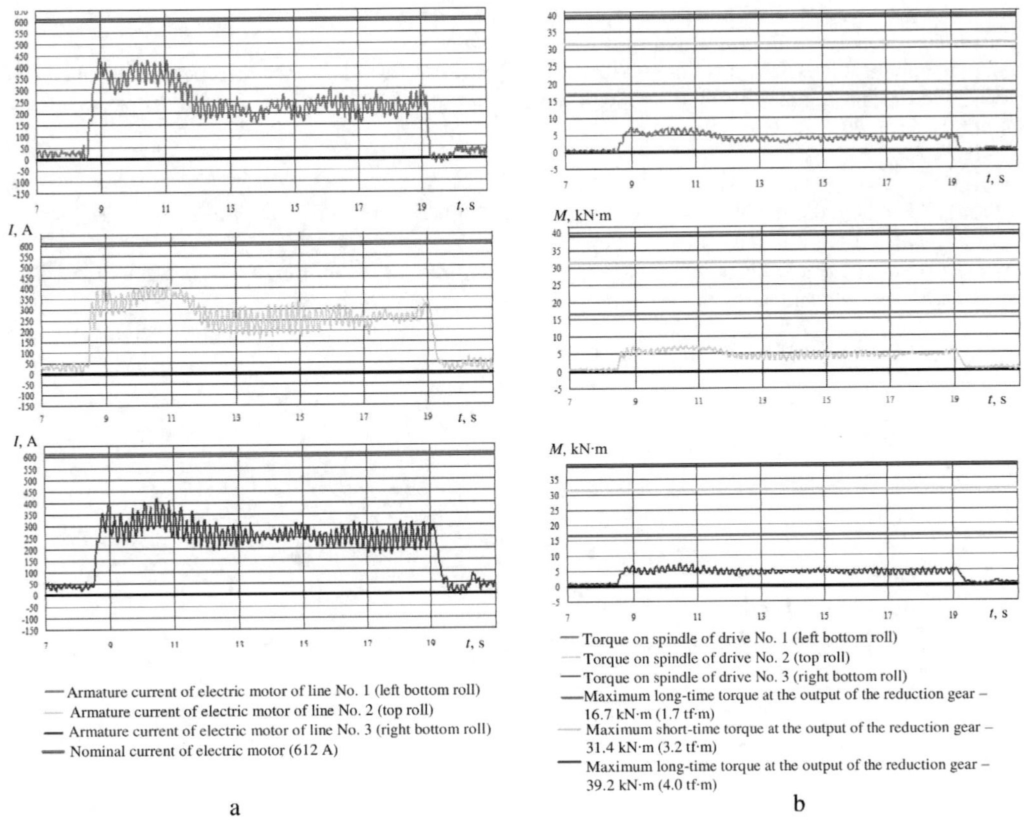

Figure 3. Oscillograms of the electric motors' currents *I* and torques on the reeling mill spindles *M* during sizing of pipe workpieces in the process of production of 146 mm diameter pipes with a wall thickness of 7.0 mm of steel grade 30G2: (a) – oscillograms of the electric motors' drives; (b) – oscillograms of the torques on the spindles.

2. For the two rolling routes under investigation the highest torques were recorded during production of 178 diameter pipes with a wall thickness of 9.2 mm of steel grade 37G2F. The maximum recorded level of torques 12 kN·m, which leads to loading of the reduction gear by long-time and short-time torque at 71.9% and 38.2%, and to loading of the spindles at 30.6 %. The total average torque on the spindles amounts to 15.5 kN·m. This leads to loading of the reduction gear by long-time and short-time torque at 73.9% and 28.7% respectively, and of the spindles – at 23.0%. The motor couplings transmit torques not exceeding 50% of the maximum permissible level. The load by torques is even less for the rolling route for manufacture of 146 diameter pipes with a wall thickness of 7.7 mm of steel grade 30G2. The maximum torque recorded during the investigation for this rolling route was 7.0 kN·m, with the total average value of 10.3 kN·m.

3. There have been little to no changes in the current loads of the electric motors when switching to manufacture of larger diameter pipes – the current loads have amounted to up to 75% of the nominal current. This is explained by the fact that the less intensive process of rolling 146 diameter pipes with a wall thickness of 7.7 mm of steel grade 30G2 is performed in the second zone of the electric motor speed control, which is characterized by power levelling and current increase with increase in speed.

4. The increase in the torques in the rolls' drivelines when switching to rolling 178 mm diameter pipes with a wall thickness of 9.2 mm of steel grade 37G2F versus process loads during rolling of pipes of conventional product range (Ø 146 x 7.7 mm of steel grade 30G2) is equal to 53%, with only 10% due to the increased strain resistance of steel, the remaining portion being due to the increase in diameter.

IOP Conf. Series: Materials Science and Engineering **718** (2020) 012002 doi:10.1088/1757-899X/718/1/012002

Conclusions

The general conclusion is that switching to rolling of extended product mix at pipe-rolling plant-140 is feasible, since despite the growth of the process loads, they do not exceed the threshold levels. Moreover, there is a margin of 20%.

References

[1] Osadchiy V Ya, Vavilin A S, Zimovets V G and Kolikov A P 2001 *Pipe production technology and equipment* (Moscow: Interment Engineering) p 608

[2] Raskatov E Yu and Spiridonov V A 2015 *Foundations of scientific researches and modelling of metallurgical machines* (Ekaterinburg: Ural University Press) p 468

[3] Konovalov Yu V, A L Ostapenko and Ponomarev V I 1986 *Calculation of sheet rolling parameters* (Moscow: Metallurgy) p 430

Research of Load Distribution Between Working and Backup Rolls of Sheet Levelling Machine with Regard to Gaps and Heat Strain

S Yu Belyaev[1], Yu B Chechulin[1], A A Fedulov[1] and M Yu Bulganina[1]

[1] Ural Federal University, 19 Mira street, Ekaterinburg, 620002, Russia

E-mail: belyaev_s_66@mail.ru

Abstract. The article addresses to distribution of loading between working and backup rolls of sheet levelling machines taking into account presence of initial gaps between them and heating that occurs during the straightening process, which is especially important in case of the high-temperature straightening. For levelling machines with a single row of backup rolls calculated dependencies were obtained to evaluate loads suffered by backup roll taking into account factors under this study and the analysis of results for the 9-roll straightening machine were carried out. The analysis showed a significant overload suffered by backup rolls.

Levelling is the final deforming operation in the process of rolled steel sheets production. It is straightening that ensures elimination of the key defects of steel sheets – wavy edges, center buckles, camber. Operating practice of levelling machines at numerous iron making and non-ferrous metallurgy plants demonstrates that setting of roll-type levelling machines (hereinafter – RLM) is quite a labor-intensive procedure [1], since quality of the straightening process and subsequently quality of the finished product depends on many factors, part of which are rather variable and impossible to control in an unbiased manner what makes them hard to take into account. The most relevant factors include the initial gap between the backup and working rolls and roll temperature values.

This challenge bears relevance especially for levelling machines intended for straightening of steel sheets under small series production conditions, which are typical for the plants manufacturing sheets from specialized steel grades or non-ferrous metals. For such plants, setting of RLM at a high level of quality considering unstable operating conditions of the rolls in terms of temperature is quite a challenging task. It takes to make certain corrections involving continuous fine adjustment or to apply additional straightening to the product. The mass and large series production of steel sheets demonstrate that the indicated factors of instability have certain impact only during the initial period of straightening of the sheet batch; the temperature condition of the roll operation becomes stable afterwards.

Therefore, the present research is aimed at identifying the heat strains of rolls during sheets straightening and evaluating the load distribution between working and backup rolls of RLM being exposed to straightening force.

The present research set a model of levelling machine with one row of backup rolls and their coaxial arrangement with the working rolls as its subject (Figure 1 (a)).

A 9-roll straightening machine designed by SKMZ (Starokramatorskii machine-building plant) was taken as a prototype of this analytical model; its rolls are D = 400 mm in diameter and the mandrel is

2000 mm long. The roll pitch of RLM under research makes 460 mm and the distance between centers of working roll supports – l = 2800 mm.

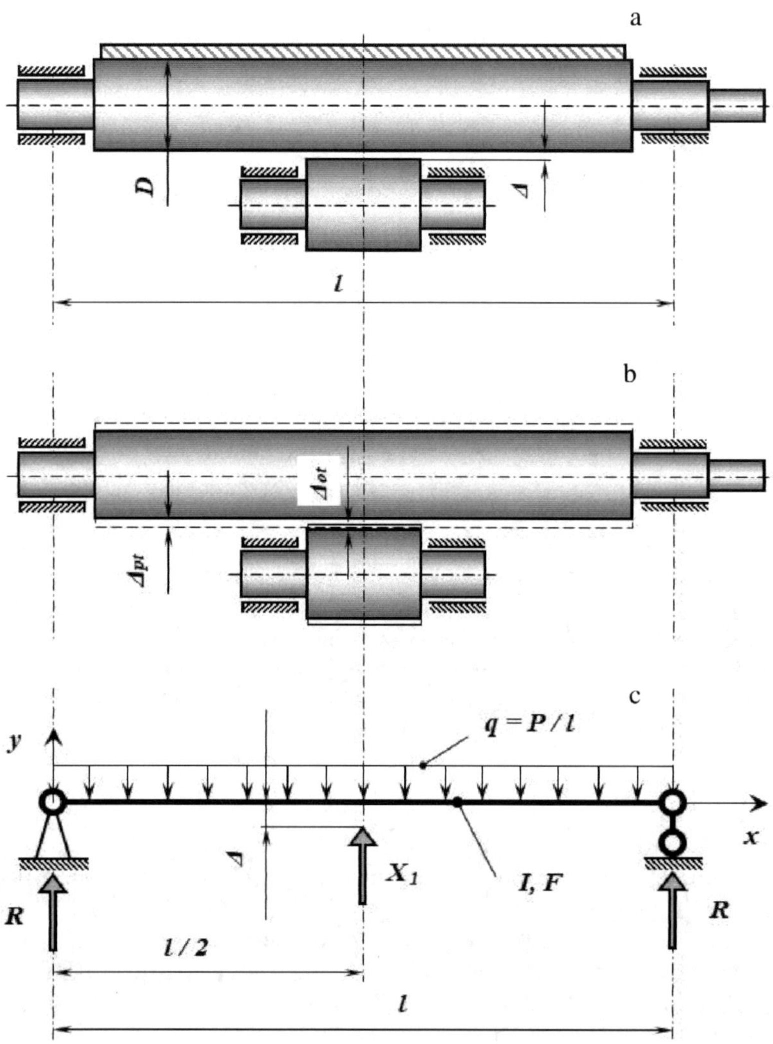

Figure 1. Design and analytical models for the working and backup roll system of sheet straightening machine: (a) – design model of the roll system; (b) – formation of heat strain; (c) – analytical model of the roll system.

The backup rolls D_{br} = 450 mm in diameter are installed in one row coaxially with the working rolls, excluding the first one and the last one. This RLM is intended for straightening of sheets up to 22 mm thick, up to 1800 mm wide made of alloys with deformation resistance up to 500 MPa. For the conditions described above, the maximum force at the 3rd roll which is exposed to the maximum load shall amount to 1900 kN.

There is a well known classical solution for the force distribution at loading of continuous beam by method of three torques [2]. This solution suggests that the backup roll should be exposed to 5/8 = 0.625 or 62.5% of the straightening force applied to one working roll. At the same time, this solution doesn't consider gaps between the rolls and possibility of their heat strain.

The total Δ between the working and backup rolls during operation (Figure 1 (a) and (b)) is defined by the algebraic sum of the initial gap Δ_0 and the total heat strain of both rolls Δ_t:

$$\Delta = \Delta_0 - \Delta_t \,. \tag{1}$$

The initial gap Δ_0 is defined by the total value of mounting clearance, value of the rolls re-grinding and their wear. As a rule the complete initial gap amounts up to 0.1-0.3 mm.

The total heat strain of the working roll Δ_{pt} and the backup roll Δ_{ot} (Figure 1 (b)) can be defined using the following formula

$$\Delta_t = 0.5 D \alpha_t t_{wr} (1 + k_D k_t). \tag{2}$$

where α_t – coefficient of linear expansion, 1/degree; t_{wr} – average increase in temperature of working roll across its section, degree; $k_D = D_{br} / D$ – relative diameter of the backup roll taken as such based on design of straightening machine within 1.0-1.15 for usual RLM design and up to 2.0 – for machines with the increased backup roll diameter; $k_t = t_{br} / t_{wr}$ – relative increase in temperature of the backup roll.

In this research, the average value of temperature increase across the section t_{wr} was a variable factor taken in the range 20°-300°. For the steel grades applied for manufacture of straightening machine rolls, the values of linear expansion coefficient at the given temperature conditions according to the reference data [3] amount to:

$\alpha_t = (1.05...1.19)\cdot 10^{-5}$ 1/degree – for alloyed structural steel type 12XH2A (12HN2A);

$\alpha_t = (1.19...1.55)\cdot 10^{-5}$ 1/degree – for bearing-type structural steel type ШХ15 (WX15);

$\alpha_t = (1.10...1.30)\cdot 10^{-5}$ 1/degree – for steel intended for roll-making type 9X.

Lower values of the linear expansion coefficients relate to the temperature range 20°-100°, and higher values – to the range 20°-300°.

Temperature distribution between the working and backup rolls can be found based on experimental data obtained via measuring the roll temperature. Should such data be unavailable, one can use the analogy with temperature distribution between working and backup rolls in the 4-high stand Quarto. Calculation of the roll system deformation is known to take the temperature increase of the backup roll in relation to the working roll considering coefficient 0.25 [4], i.e. k_t = 0.25. For RLM, a recommendation to select the value of relative temperature increase of the backup roll within the range k_t = 0.2-0.3 is feasible. Lower values k_t can be applied to the backup rolls of higher diameter in comparison to the working rolls, while higher values k_t – to RLM in which backup rolls are arranged in a staggered way.

A special calculation was made for 9-roll straightening machine of SKMZ design, which was taken earlier as a prototype so that to assess the total impact of the initial gap and heat expansion of the rolls. The calculation was made for various levels of temperature increase of the working roll within the range 20°-300° with a pitch of 20°. The relative backup roll diameter was taken based on design parameters of RLM $k_D = D_{br} / D$ =450/400=1.125, and the relative backup roll temperature increase $k_t = t_{br} / t_{wr}$ was taken as 0.25. Table 1 shows the calculation results for the rolls made of steel 12XH2A (12HN2A), with the initial gap 0.1 mm and 0.2 mm.

Negative values of the total gap between working and backup rolls means there is no gap and both rolls are preliminarily loaded without any external load/impact. According to analyses of the obtained results, increase in temperature of the working roll up to 40°-80° makes the heat strain of the roll system compensate the initial gap Δ_0, and force distribution between working and backup rolls is nearly equal to a theoretical value. Increase in temperature of the working roll over 100° causes rapid growth in preliminary load of the rolls.

With the obtained data on the value of the total gap between the rolls taken into account, an analytical model for the working roll was formed, represented at Figure 1 (c). In this model the working roll is represented as a simple beam with length l round in section, with the area F and inertia moment I. The beam is loaded with an equally distributed load $=P / l$, where P – strip levelling force applied to the working roll. The impact applied by the backup roll to the working roll is accounted by the statically indeterminate force X_1, which becomes effective only upon reducing gap Δ to zero, and reaction of the bearing support of working roll makes R.

Statically indeterminate force X_1 was determined from solution of the classical equation of the force method [5], since the gap between the rolls made it impossible to use a theorem related to three torques in this case:

IOP Conf. Series: Materials Science and Engineering **718** (2020) 012003 doi:10.1088/1757-899X/718/1/012003

Table 1. The total heat strain and total gap between working backup rolls.

Increase in working roll temperature t_{wr}, degrees	Total heat strain of the rolls Δ_t, mm	Total gap between the rolls Δ, mm with the initial gap	
		$\Delta_0 = 0.1$ mm	$\Delta_0 = 0.2$ mm
20	0.054	0.046	0.146
40	0.108	-0.008	0.092
60	0.161	-0.061	0.039
80	0.215	-0.115	-0.015
100	0.269	-0.169	-0.069
120	0.354	-0.254	-0.154
140	0.413	-0.313	-0.213
160	0.472	-0.372	-0.272
180	0.530	-0.430	-0.330
200	0.589	-0.489	-0.389
220	0.671	-0.571	-0.471
240	0.732	-0.632	-0.532
260	0.793	-0.693	-0.593
280	0.854	-0.754	-0.654
300	0.915	-0.815	-0.715

$$\delta_{11}X_1 + \Delta_{1P} = -\Delta, \tag{3}$$

from which:

$$X_1 = \frac{-\Delta - \Delta_{1P}}{\delta_{11}}, \tag{4}$$

where Δ_{1P} – displacement from application of external forces towards statically indeterminate force X_1, mm; δ_{11} – displacement from application of single force towards statically indeterminate force X_1, mm/N.

Displacement from application of single force and external force was defined using standard methods of material strength. As a result the following ratio were obtained, considering both bending torques and lateral forces:

$$\begin{aligned} \Delta_{1P} &= -\frac{5}{384}\frac{Pl^3}{EI} - k\frac{Pl}{8GF}, \\ \delta_{11} &= \frac{l^3}{48EI} + k\frac{l}{4GF}, \end{aligned} \tag{5}$$

where E, G – elasticity modulus and shear modulus accordingly, MPa; k – coefficient related to the form of cross-section during shear; for round section $k = 1.185$.

Should the calculation of displacement consider only bending torques, the represented ratio shall retain only the first summands.

Upon substitution of displacement values into the formula intended to find the statically indeterminate load and transforming, we shall obtain the following ratio:

- considering only bending torques

$$X_1 = \frac{5}{8}P - \Delta\frac{48EI}{l^3}, \tag{6}$$

- considering both - bending torques and lateral force

$$X_1 = 0.25P\frac{5\left(\frac{l}{D}\right)^2 + 6k(1+\nu)}{2\left(\frac{l}{D}\right)^2 + 3k(1+\nu)} - \Delta\frac{3\pi ED^2}{2l\left[2\left(\frac{l}{D}\right)^2 + 3k(1+\nu)\right]}, \qquad (7)$$

where ν – coefficient of lateral deformation (Poisson's coefficient).

For the steel grades used for manufacture of straightening machine rolls, values of mechanical properties according to the reference data [3, 5] at temperature range 20° up to 300°C are as follows:

- Poisson's ratio $\nu = 0.28\ldots0.30$;
- elasticity modulus $E = (2.11\ldots1.90)\cdot10^5$ MPa.

Should the gap between the rolls be equal nil (provided that we consider only bending torques), one shall obtain the force at the backup roll equal to 0.625 from the straightening force at the working roll, which gives a clear correspondence with the above mentioned classical solution of this task [2]. In case we consider both - bending torques and lateral forces for the 9-roll straightening machine under the study under the similar conditions, the proportion of complete straightening force applied to the backup roll shall amount to 0.619, which is quite similar to the well known solution (error not more than 1%).

The load suffered by the backup roll is convenient to calculate as a relative value instead of absolute force values – in certain proportion from complete force applied to the working roll, through the ratio X_1 / P. As such, the relative force at the backup roll in its extended form shall amount to:

- considering only bending torques

$$\frac{X_1}{P} = 0.625 - \frac{48EI}{Pl^3}\left[\Delta_0 - 0.5D\alpha_t t_P\left(1 + k_D k_t\right)\right], \qquad (8)$$

- considering both bending torques and lateral forces

$$\frac{X_1}{P} = \frac{0.25}{2\lambda^2 + 3k(1+\nu)}\left\{5\lambda^2 + 6k(1+\nu) - \frac{6\pi ED^2}{Pl}\left[\Delta_0 - 0.5D\alpha_t t_{wr}\left(1 + k_D k_t\right)\right]\right\}, \qquad (9)$$

where $\lambda = l / D$ – relative length of the roll.

The overall estimation of calculation methods with regard to heat expansion of rolls for the 9-roll RLM took making of the following graphs at Figure 2 presenting the relative force at the backup roll, as a function of temperature increase of the working roll with the initial gap 0.1 mm.

According to analyses of the represented graphs, increase in temperature of the working roll up to 50°C results in quite insufficient gap value (preliminary loading) between the rolls, and the results obtained by application of different approaches to solution of this task are relatively equal being in compliance with a well-known classic solution amounting to 0.625. Increase in temperature over 50°C leads to considerable force increase applied to the backup roll, caused by effect of bending torques and lateral forces, i.e. through the ratio taking into account complete strain energy. This effect can be explained by the fact that together with the increase in load suffered by the backup roll, a simultaneous reduction in reaction at the bearing supports of the backup rolls takes place, which also causes subsequent reduction in the share of bending torques in the total strain energy value. The total straightening force is applied to the backup roll, when the working roll temperature is increased by 150° C, and its overload in comparison to a theoretic solution makes 60%. Should the temperature of the working roll increase up to 300° C, the backup roll shall be exposed to force which is almost 1.5 times higher than the complete maximum straightening force at the working roll, and the overload shall make 135% in comparison to a theoretic solution. Together with that reaction of bearing supports shall have same tendency as that of the process load, increasing friction torque in the roll bearings.

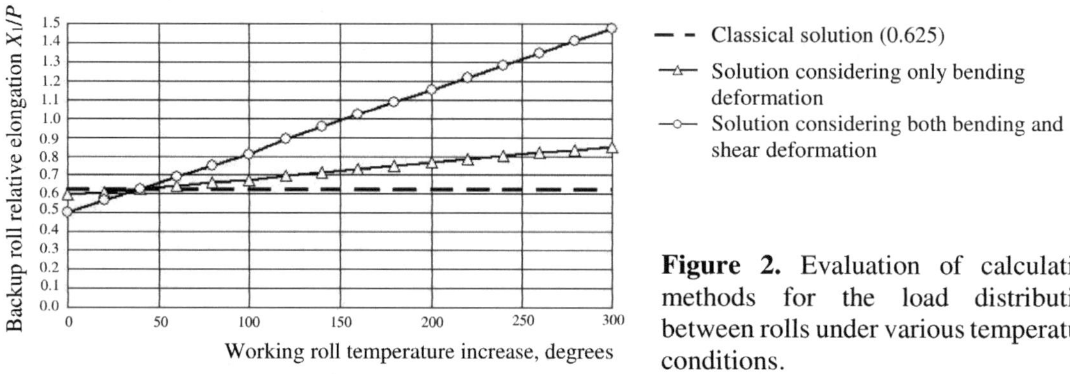

Figure 2. Evaluation of calculation methods for the load distribution between rolls under various temperature conditions.

Apart from overload suffered by the backup roll, loading of RLM roll system presented in the case above leads to change in position of the working roll, thus causing not negligible fluctuations in moving of its operating surface along the width of a sheet to be straightened. This undermines straightening quality of the product.

Conclusions

Summarizing results of the above research, it takes to draw a conclusion as follows:

1. The probable overload of backup roll as much as 2.35 times in respect to a well-known theoretical solution has been justified. The initial gap between rolls at the existing levelling machines shall be set in a way to avoid substantial backup roll overload. For the newly designed equipment it is necessary to consider probable increased estimated load suffered by the backup rolls.
2. Setting of levelling machines shall take it necessary to consider not only total displacement of the crossbeam with the rolls installed on it, but also gaps and heat strain of the rolls.

References

[1] Nedorezov I V 2003 *Modelling of metal levelling processes at the roll-type levelling machines* (Ekaterinburg: AQUA-PRESS) p 256

[2] Korolev A A 1985 *Structure and design of machinery and devices of rolling mills* (Moscow: Metallurgy) p 376

[3] Sorokin V G and Gervasiev M A 2003 *Steels and alloys. Steel grades guide* (Moscow: Intermet Engineering) p 608

[4] Parshin V S, Kostrov V P, Somov B S and at al 2004 *Machinery and facilities used for processing of non-ferrous metals and alloys* (Yekaterinburg: GOU VPO UGTU-UPI) p 383

[5] Pisarenko G S, Yakovlev A P and Matveev V V 1988 *Reference guide for the strength of materials* (Kiev: Nauk. dumka) p 736

IOP Conf. Series: Materials Science and Engineering **718** (2020) 012004 doi:10.1088/1757-899X/718/1/012004

Theoretical Research on the Pipe Rolling Process in Order to Determine the Deformation of the Pipe Billet and the Load on the Rolling Tool

Yu B Chechulin[1], A A Fedulov[1], A I Dronov [2], V A Morgunov[2] and M. Yu. Bulganina[1]

[1] Ural Federal University, 19 Mira street, Ekaterinburg, 620002, Russia
[2] Pervouralsk New Pipe Plant, 1 Torgovaya street, Pervouralsk, Sverdlovsk region, 620143, Russia

E-mail: a.a.fedulov@urfu.ru

Abstract. The technology of the shell case rolling into the rough pipe on an automatic mill of the pipe-rolling plant PRP 220 was considered. In case of the pipe rear end stability losing for a special thin-walled assortment of rolling products the theoretical research was carried out based on the FEM. As the result of the research the nature of such flaw as a "fin" on the end of the pipe was investigated, the evaluation of the loading on the automatic mill for the PRP 220 was obtained for different process parameters.

Hot rolling of wide assortment of seamless pipes made of carbon, alloy and high alloy steels on pipe-rolling plants (PRP) with an automatic mill includes the following sequence of operations: furnace heating of a round billet, piercing, rolling on a short mandrel through an automatic mill, rolling through reeling machines, calibration or reduction and flattening [1]. The most important operation is rolling of the shell into a pipe with a given wall thickness on an automatic mill [2].

Rolling of the pipes through an automatic mill with a ratio of the outer diameter to the wall thickness of more than twenty five is accompanied by the formation of a defect at the end of the pipe in the form of a longitudinal fold [3]. This fold, shaped like a fin, is shown in Figure 1 (a). In addition, significant destruction of the end sections occurs (Figure 1 (b)). The consequence of this form change of the billet is an increase in the load on the working tool of the automatic mill and reeling machines, as well as their drivelines.

In order to analyze the shape change of the pipe billet when rolling through an automatic mill, three-dimensional modeling of the process was performed with a help of FEM software [4]. The calculation scheme of the mathematical model includes a viscous-plastic billet (pipe shell), absolutely rigid work rolls and a mandrel.

The calculation was made starting from the moment when the billet entered the rolls' opening and up until it was completely out of the rolls. As you know, rolling through an automatic mill is carried out in at least two passes. In this regard, the calculation was performed for two operations. For the second rolling operation through an automatic mill, the billet obtained after the first pass was moved to the input side of the rolls and turned over to an angle of up to 90°. Then, similarly to the previous calculation, the second pass was modeled in rolls of the same profile with the mandrel replaced.

a

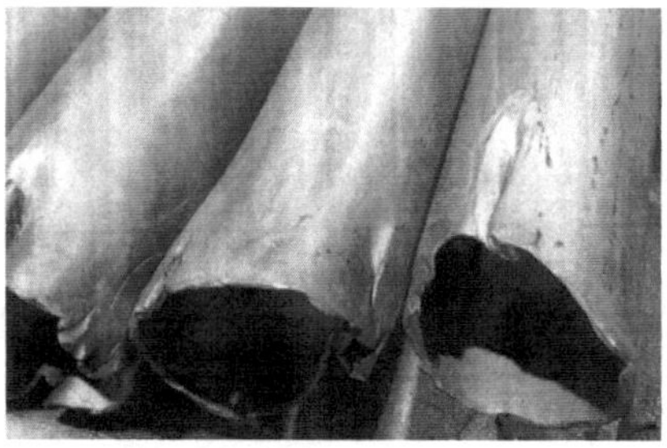

b

Figure 1. Defects of end sections of thin-walled pipes after rolling through an automatic mill (a) and before flattening (b).

The performed theoretical study made it possible to observe a loss in the lateral stability of the end of the pipe when rolling, which begins with ovalization of the cross section of the billet, followed by the transition of metal into the roll gap and rolling of the metal into the “fin” (Figure 2). At the same time, the narrowed section pipe is put on the mandrel.

For a different assortment of pipes, the peak at the time the pipe exits the stand decreases with the increase of wall thickness. For thin-walled pipes in the indicated range, the peak goes up to 27.6% (of the average rolling force) and gradually smooths out. The graph of axial forces is also of a similar nature, which is also accompanied by a characteristic peak at the moment of “wrinkling” at the end of the pipe (14.2-22.9% of the average values of axial force for various ratios of the outer diameter of the pipe to the wall thickness).

During the second pass, depending on the tilt angle of the billet, an additional deformation is added to the existing loss of stability of the end of the pipe when it leaves the stand, due to the need to roll out the “fins” formed during the first pass. In the real rolling process, in the places of rolled “fins”, destruction of the pipe end appears (Figure 1 (b)).

Figure 3 shows graphs of the calculated forces taking place on the work rolls and the mandrel during the rolling process. At the same time, a new pair of "fins" is formed due to the transition of metal into the roll gap, which subsequently impedes the normal rolling process in reeling machines, significantly (up to three times) increasing the rolling force in the final part of the rolling.

To establish the relation between the size of the “fins” resulting from the second pass and the metal transition in the deformation zone in the transverse direction, a theoretical study examined various tilting angles of the billet between the passes.

The graphs of the forces taking place on the working tool during the second pass are shown in Figure 4.

It was established that this defect takes place in all cases, regardless of the tilt angle. The highest values of the force are achieved at a 90° tilting angle due to a larger reduction of the pipe wall in the rolls’ opening (the difference between the maximum average force at a tilting angle of 0° and 90° is 36.3%).

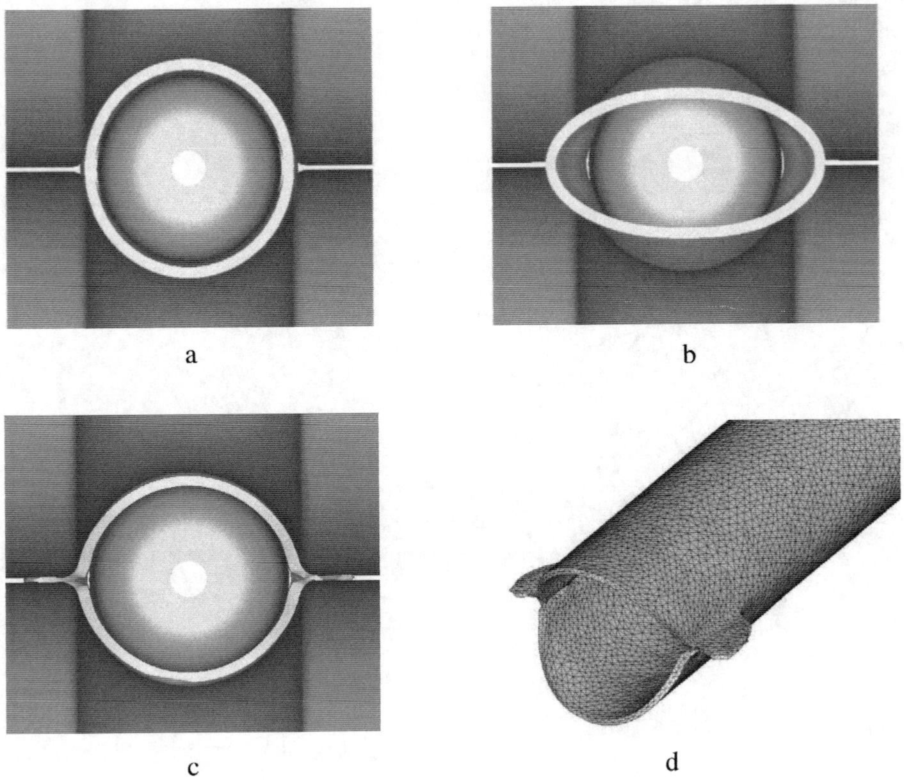

Figure 2. The end of the billet when rolling (a-c) the stages of stability loss of the billet end at the end of rolling (d) final form-change of the billet.

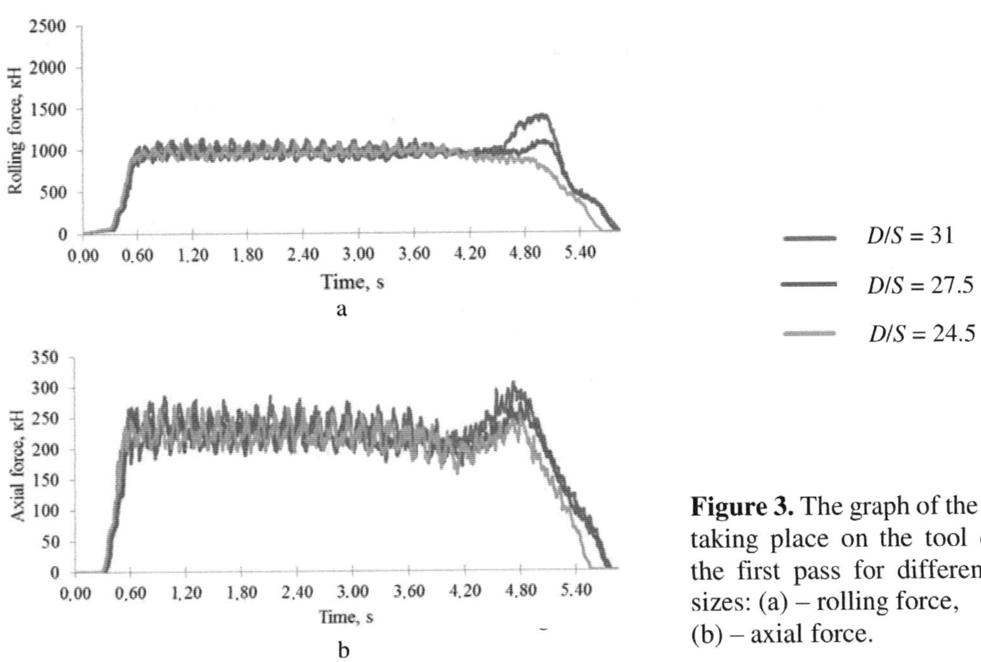

Figure 3. The graph of the forces taking place on the tool during the first pass for different pipe sizes: (a) – rolling force, (b) – axial force.

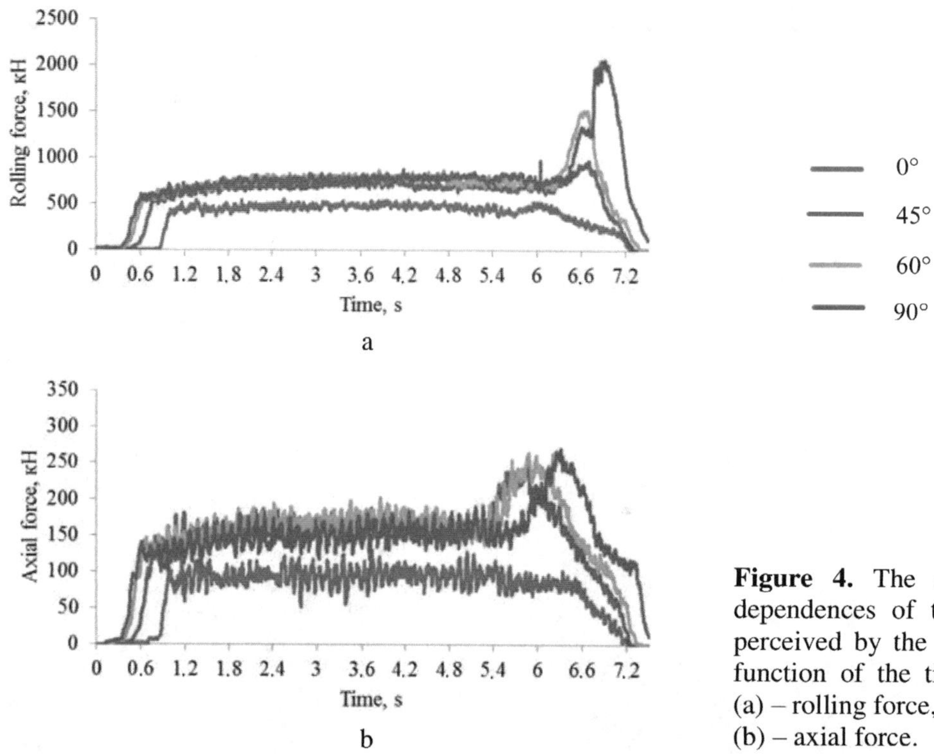

Figure 4. The calculated dependences of the forces perceived by the tool, as a function of the tilt angles: (a) – rolling force, (b) – axial force.

As can be seen from the graphs, the moment of rolling of the "fins" is accompanied by a ramp in the rolling force, which in turn is accompanied by an increase in axial force due to the putting of the pipe billet on the mandrel. The maximum ramp of the force is observed at a tilt angle of 90°, at which its value exceeds the average value of the rolling force by 2.78 times and the average value of the axial force by 1.80 times.

It should be noted that the minimum tilting angles increase the ovalization of the pipe (ovalization coefficient from 1.0 when tilting 90° to 1.04 when tilting is absent) and walls' thickness variation in the cross-section (from 4.8% at tilting angle 90° to 30% when there is no angle).

As practical tests have shown, the elimination of the formation of "wrinkles" in the form of "fins" at the end sections of thin-walled pipes with a diameter to wall thickness ratio of more than twenty five can be achieved by limiting the ovalization of these sections.

Conclusions

As a result of modeling the process of rolling the shell through the automatic mill of the pipe-rolling plant, it was found that the end of the pipe when rolling, due to reduced hardness, loses stability and ovalizes. This is accompanied by the transition of pipe metal into the roll gap and the formation of a pipe defect in the form of a "fin". In this case the load on the working tool has stick-slip nature, and the forces significantly exceed their average values on the working tool during the rolling process.

Practical tests of additional measures to reduce ovalization of the end sections showed the possibility of eliminating the indicated defect and reducing the expenditure coefficient.

References

[1] Osadchy V Ya, Vavilin A S, Zimovets V G and Kolikov A P 2001 *Technologies and equipment for pipe production* (Moscow: Intermet Engineering) p 608.

[2] Osadchy V Ya and Kolikov A P 2012 *Productions and quality of steel pipes* (Moscow: Moscow State University of Instrument Engineering and Computer Science) p 370

[3] Kolikov A P 1998 *Machines and equipment for pipe production* (Moscow: National University of Science and Technology MISiS) p 536
[4] Karamyshev A P, Nekrasov I I, Parshin V S, Fedulov A A and Pugin A I 2012 *Metallurg* **2** 53–55.

IOP Conf. Series: Materials Science and Engineering **718** (2020) 012005 doi:10.1088/1757-899X/718/1/012005

Principles of Coal Blend Formation Required to Obtain Coke with Target Characteristics of Coke Strength after Reaction (CSR)

E V Karunova[1,2], **A A Kalko**[1], **E A Volkov**[1], **and O A Kalko**[2]

[1] PAO Severstal, 30 Mira street, Cherepovets, 162608, Russia
[2] Cherepovets State University, 5 Lunacharsky street, Cherepovets, 162600, Russia

E-mail: evkarunova@severstal.com

Abstract. A model has been developed that allows formation of a coal blend for production of coke with the specified characteristics of CSR in industrial conditions. The model is based on experimental data on the CRI and CSR values of coal concentrates obtained during individual coking using a SKW Karbolab plant.

Nowadays the coke quality is globally assessed by a number of values characterizing its properties in regards to its functions performed in certain areas of the blast furnace. For example, requirements of European blast furnace metallurgists to coke quality are as follows: mechanical strength (M_{10} and M_{40}), grain size (contents of coke class > 80 mm, and in skip coke – class < 40 and < 20 mm), the values CSR (coke strength after reaction) and CRI (coke reactivity index) and chemical contents [1].

The assessment method of coke «hot» strength (CSR) proposed by the Japanese company «Nippon Steel» in the 1970s (further NSC) has become a regular practice of both, domestic and foreign metallurgy, since the above mentioned values CRI and CSR complete the standard parameters of coke mechanical strength control to the utmost and adequately display changes in quality of coal blend and technological factors of coke and by-product process [2, 3].

Despite being wide spread in the global and domestic practice, the NSC method has a number of relevant drawbacks, the main of which are low reproducibility, high cost of equipment and long duration of analysis performance [4]. This means that coke manufacturers are unable to react quickly to changes in raw material and (or) technological factors. This makes usage of mathematical models of coke quality forecast in terms of CRI and CSR quite reasonable.

Development of mathematic models and theoretical basics for the forecasting methods of physical and mechanical properties of coke using coal property values has been for many years one of the key problematic issues of coke production, since the solution of this issue is of high relevance in terms of application. Most of iron and steel-making plants possess their own adopted and practice-proven forecast formulas for the «cold» coke strength values. As a rule, those are based at the coal process characteristics demonstrated during its thermal conversion: properties of plastic mass, its caking ability and data on the coal genetic nature, coking technology and coke quenching method [5-6].

A number of mathematic models [7-10] are introduced presently to enable forecasting of CRI and CSR values of coke. Such models are based on the ash basicity index and chemical and petrographic properties of coal blends taking into account methods of their preparation, coking period and quenching method. Such models usually rest upon bulk process data sets of a plant taken for a certain period or results of laboratory-based and box coking. The key drawback of those models is limitation of their application (the model gives a valuable forecast only for the coke quality values at certain raw material base of coals).

Coal blend formation at coke and by-product plants is mostly performed by practical knowledge based on experimental coking (laboratory-based or box coking). Lack of flexibility under the changeable conditions at the raw material market is a substantial weak point of this approach to coal blend formation. That's why one of the relevant issues for PAO Severstal is to select ratio of the coal concentrates using mathematic models, theoretical background and forecasting methods of physical and mechanical properties of coke with regard to coal properties. Settlement of this issue shall enable minimizing the risks arising from introduction of new suppliers and shall allow to take adequate solutions in terms of ratio between net cost and quality of coal concentrates. It should be noted that such research work is carried out by OAO Altai koks together with AO Eastern Research coal chemistry institute [11].

Control and forecast of coke quality in the iron-making plant of PAO Severstal is performed using laboratory facility type SKW «Karbolab» made by the Radiotechnical Institute of Poland (Figure 1).

Figure 1. Laboratory coking facility.

Usage of this laboratory oven allows carrying out of experimental coking of coals constituting the raw material base of CherMK, both actual and artificially modeled coal blends under conditions simulating the industrial coking process. An important thing about this laboratory facility is the stability of coking parameters, which makes it possible to evaluate the impact and to match the properties of the initial raw materials and properties of the produced coke. Moreover adoption of this facility reduces the price of the high cost semi-industrial and industrial experimental coking procedure in the operating coke ovens.

Research work performed by this laboratory oven allows to:

- deepen the knowledge of process value degree of certain coal concentrates;
- improve the ranking method for the coal concentrates which constitute the coking base at PAO Severstal;
- evaluate the influence of individual factors on the CSR value under conditions of laboratory;
- develop a method assisting to find an adequate balance of the incoming coal concentrates so that to receive the target CSR value to be used in iron-making plant.

During development of a method of coal blend matching for production of coke with the target characteristics of CSR, we faced a necessity to check the assumption of additivity of the coke quality values obtained at individual coking procedures of separate coal blend components, and quality values obtained during coal blend coking. Should the data possess additivity, there arises a possibility to use the results of laboratory-based coking of individual coal concentrates for forecasting of the industrial coke quality.

In this research work, we have addressed to a bulk data set incorporating individual laboratory-based coking procedures of 120 samples of pure coal concentrates and 35 samples of artificially modeled coal blends made from those coal concentrates.

At Figure 2 you can see the interdependence between values of coke strength after reaction with carbon dioxide received with the coal blends and calculated value determined through the portion of

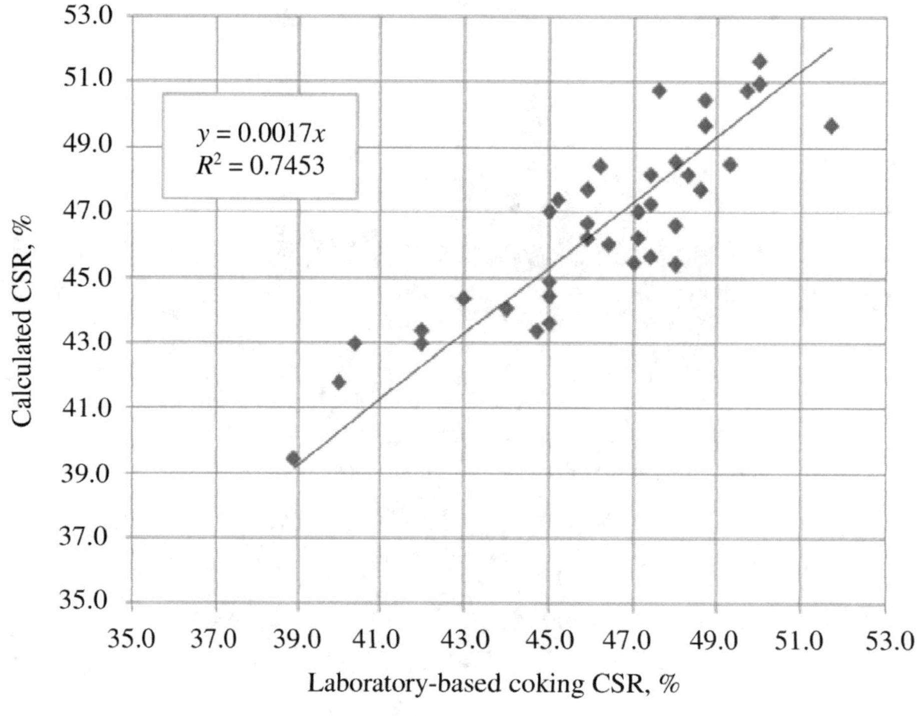

Figure 2. Interdependence between strength characteristics of coke obtained during laboratory-based coking of coal blends and the calculated value based on the portion of involvement in the blend and individual CSR value of each blend component.

individual concentrates involvement in the coal blend intended to be coked. The confirmed additivity of individual coking values allowed their inclusion as the key value of mathematic model of coal blend trial during development of patent RU2608524C1 dd 19.01.17.

Thus we can state that for the coal concentrates constituting a raw material base of the plant individual laboratory-based coking is practicable. Coke produced that way is then analysed all through; the results of the set of analyses can be used for quality evaluation of coke produced through the process flow. The key value to be identified in the produced coke samples is CSR.

Thus coal blend is formed based on the data on CSR, individual coal concentrates taking into account additivity of those values. Coal blend formation shall also address to the blast furnace process demand maintaining the certain range of strength characteristics of oven coke. Theoretical and industrial values are calculated using reduction factors via linear dependence [12].

At Figure 3 you can see a graph outlining the interdependence between coke strength value after reaction with carbon dioxide (CSR), calculated through individual coking procedures using the forecast formula and actual average value taken over the period when the coal blend consumption charts have been used at the iron-making plant of PAO Severstal.

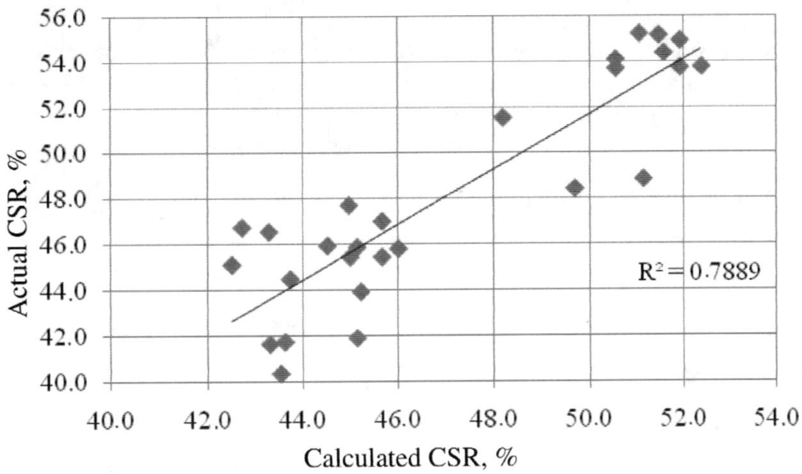

Figure3. Interdependence between calculated and actual values of coke strength CSR for coke division of iron-making plant of PAO Severstal.

Deviations of shift-wise actual values of CSR from the target values set in the al blend consumption charts have been analyzed in order to evaluate deviation between forecast and actual values of coke strength after reaction with carbon dioxide (CSR). It has been found that 85% of all the obtained data lies within the range of ±3%, which is in conformity with the repeatability limit used for determination of CSR. At Figure 4 you can see a diagram of deviation breakdown.

It should be noted that the developed model is used with success in the coke and by-product process stage at PAO Severstal. Its implementation enabled achievement of the *CRI* and *CSR* values of the produced coke equal to 25-35% and 48-60% accordingly which is close to the adequate values [13] and in its average meets the requirements for coke characteristics imposed at the plants of Western Europe [14].

Conclusions

Based on the findings of performed research work one can conclude the following:

1) values of individual coal coking are additive;

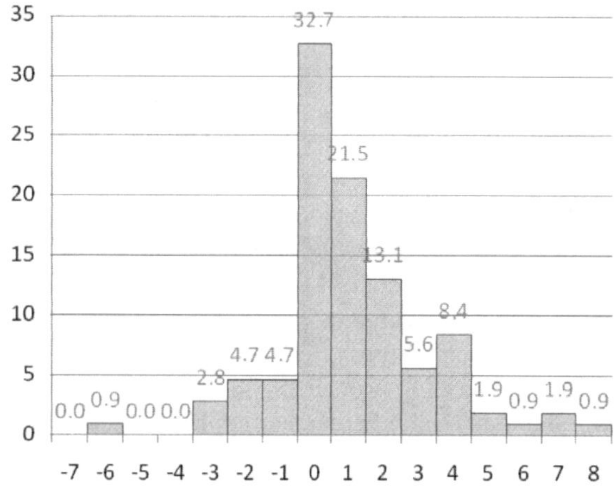

Figure 4. Diagram of deviations breakdown of actual values of CSR from the target values as per coal blend consumption chart.

2) strength of coke acquired from a blend of coal concentrates after reaction with carbon dioxide can be calculated based on the portion of coal concentrate involvement in the coal blend to be coked and CSR values achieved during individual coking;

3) application of the proposed model may allow forecasting of the coke quality both for «cold» strength and CSR;

4) selection of the adequate contents of coal blend to be coked taking into account all the coke characteristics relevant for blast furnace procedures has become possible.

References

[1] Zolotukhin Yu A, Osadchii S P, Denisenko E V, Andreichikov N S, Kuprygin V V and Zorin M 2019 *Coke and chemistry* **5** 12–37

[2] Muchnik D A 2010 *Coke and chemistry* **10** 17–23

[3] Gurkin M A, Vinogradov E N, Volkov E A, Karunova E V and Kalko A A 2017 *Method of coal blend formation used to obtain metallurgy coke with the target CSR value* (Patent RU2608524C1)

[4] Ukhmylova G S 1995 *Coke and chemistry* **4** 37–41

[5] Geerdes M, Chenio R, Kurunov I, Lingardi O and Rickets D 2016 *Modern blast furnace procedure. Introduction* (Moscow: Metallurgizdat) pp 69–70

[6] Stukov M I, Kogan L A, Kosogorov S A and Kukolev Ya B 2011 *Coke and chemistry* **3** 50–52

[7] Stepanov Yu V, Popova N K and Koshkarov D A 2005 *Coke and* chemistry **5** 26–34

[8] Gabov A I, Trifanov V N, Konovalova J V and Karunova E V 2012 *Coal blend formation method used to obtain metallurgy coke* (Patent RU2461602)

[9] Stankevich A S and etc. 2000 *Coke and chemistry* **5** 2–10

[10] Bulanov E A, Zainutdinov V N, Kuznetsov V Y and Zinovieva L A 2005 *Coke and chemistry* **5** 23–26

[11] Savchuk N A and Kurunov I F 2000 *Blast furnace plant at the turn of XXIst century* (Moscow: Chermetinform) p 42

[12] Bazegskii A E, Ryabichenko A D, Khamidulin F Z, Pianov B F and Stankevich A S 2002 *Coke and chemistry* **9** 15–22

[13] Stankevich A S, Guilliazetdinov R R, Popova N K and Koshkarov D A. 2008 *Coke and chemistry* **9** 37–44.

[14] Lialiuk V P, Sokolova V P, Schmeltzer E O, Timofeeva D Yu and Bereza V V 2014 *Coke and chemistry* **6** 15–22.

IOP Conf. Series: Materials Science and Engineering **718** (2020) 012006 doi:10.1088/1757-899X/718/1/012006

Problems and Peculiar Aspects of Training of the Future Engineers in Relation to Higher Education System Remodelling

A V Kozhevnikov[1]

[1] Cherepovets State University, 5 Lunacharsky street, Cherepovets, 162600, Russia

E-mail: avkozhevnikov@chsu.ru

Abstract. The article attempts to analyze the current situation with University education, proposes a model of a modern University and training of engineers.

1. Introduction

Higher engineering education, same as humanities and teacher education in Russia is currently suffering certain indefiniteness. On the one hand strive for enabling the remodeling of the higher education made us try all the existing Western practices related to organization of training process for the future engineers, to the extent of coming of MIT professors to Skoltech, and various futurologists keep saying that the next 15 years shall see a fundamental change in education effected by digital technologies and the coming industrial revolution [1]. On the other hand business, particularly iron and steel making companies point out that existing professional training is insufficient in terms of challenging the current industrial tasks, that Universities are out of touch with industrial production, that remodeling is a critical issue of today. Still neither business, nor futurologists can indicate the exact ways to make the present training of engineers at the University meet the necessities and realities of the day.

2. Problematics

Yes, indeed we need to confess that higher engineering education is facing the hard times. Here are some problems of the modern Universities which assist cultivating their existing image in the eyes of business:

1. *Mass nature of higher education*. The Universities are no longer postured in the society as a place for upbringing the future elite, but merely represent a place for rendering educational services to those who have completed 11 grades in order to get the University diploma afterwards.

2. The second reason resulting from the first mentioned above – *motivation* of University applicants and students. I know from my own experience of work at the University that the portion of motivated students (those who realize what people expect from them and what they need to do at the University) makes about 10-20% from the total number of students. Most of the University students if I may say so «simplify their study on their way to get a document certifying their qualification to the maximum extent».

Half of those students are enrolled to the University by chance, but this is not the reason. This is caused by the existing training formats (see par. 3 below).

IOP Conf. Series: Materials Science and Engineering **718** (2020) 012006 doi:10.1088/1757-899X/718/1/012006

3. *The existing training formats.* Lecturing, narrating of the textbooks contents to the students, courses with unclear intention in terms of competency map forming taken by students – all that being rendered over 4 years of bachelor's degree course and 2 years of master degree course brings down the interest of students in their study. The information available in the network replaces the teacher as a knowledge translater. Educational programs are mostly not flexible, suggesting merely formal selection of elective courses. One can compare a typical University pathway with a pipe: on one end you enter the University training program, on the other end you come out with a document certifying your higher education. Training programs are designed to cover the competencies of the existing academic staff instead of competencies of the future specialists required by business. In this regard what underlies training is the maxim «We teach you what we know, but not what you need».

The existing educational formats facilitate established opinion that University study is a waste of time, which can be proved by the latest statistics for reduction of popularity of the Universities among the school children.

4. Presence of *inertia* while training the human resources with higher education in the existing educational system «bachelor's degree – master's degree» does not allow for prompt response for the business requirements. Remodeling and upgrading of educational programs under these conditions may give its effect at least after the period of 2-4 years.

5. Certain *asociality of «educational eco-system»* in domestic Universities. This proves to be one of the key aspects which differs Russian education from West European or American one. For about 20 years already the Universities in the USA and Western Europe have developed and adopted interactive means of training, control, organization and implementation of education and research processes based on Internet technologies, the main objective of which is *on-line* cooperation, access of the students to *big data* and creation of a global community with network education based at the best competencies of the leading global Universities [2, 3].

This issue of educational eco-systems' asociality, development and adoption of Internet technologies and artificial intelligence in the system of education is currently being dealt with by a project of National Technology Initiative (NTI).

All the indicated and other existing problems as well as realizing that "something is wrong" with the higher education lately have brought forth many discussions in professional society related to the topics so popular today: digital transformation at the Universities, project activity in higher education, *soft or hard skills*, professions and competencies of the future, *big data*, individual nature of education etc.

However in chase of those popular trends such discussions sometimes bring to life theses dangerous for business development, for example:

- «future belongs to online education»: it is absolutely clear that none of potential employers in the sphere of engineering would like to hire specialists trained in a remote manner;
- «bachelor's degree course does not cover training for qualification, rather some initial general degree of higher education and development of personality».

The situation mentioned above makes businessmen and industry-related people concern about personality and methods of training the future motivated professionals and experts with basic knowledge, able to design technologies and equipment, to bring up innovations facilitating development of industrial companies in various spheres; another issue that arises in this regard is the real mission and role of the Universities in training of professional personnel for industrial and research activities.

3. Peculiarity of the University-based education

The fundamental distinction between University-based education and that of the secondary school or secondary professional education organizations is the research process.

Research shall «penetrate» all the educational activities at the Universities (Figure 1), meaning that it is vitally important to engage Academic staff dealing with active research and expertise in their sphere to the educational process. Here however lies one unsettled discrepancy between business and

University-based training: if a researcher teaches a student what he knows himself, then he shall lose his unique competence, which is a real threat for his position at the University. That's why the University-based training shall be founded on creating of environment for independent / team research work and acquiring of new knowledge instead of drilling. Thus together with research process, creation of new knowledge instead of their repeating at the University is the second key distinctive feature of the University from other types of educational institutions.

Absence of true research activities at the University and respectively new knowledge results in a sort of discord and failure to understand the processes going at the University for eventual consumers. Without research and researchers the University becomes more like a mere secondary school, or a college; in this situation it is hard for Universities to identify themselves at the labor market as an entity able to present the results of educational process as a certain set of competencies with its graduates.

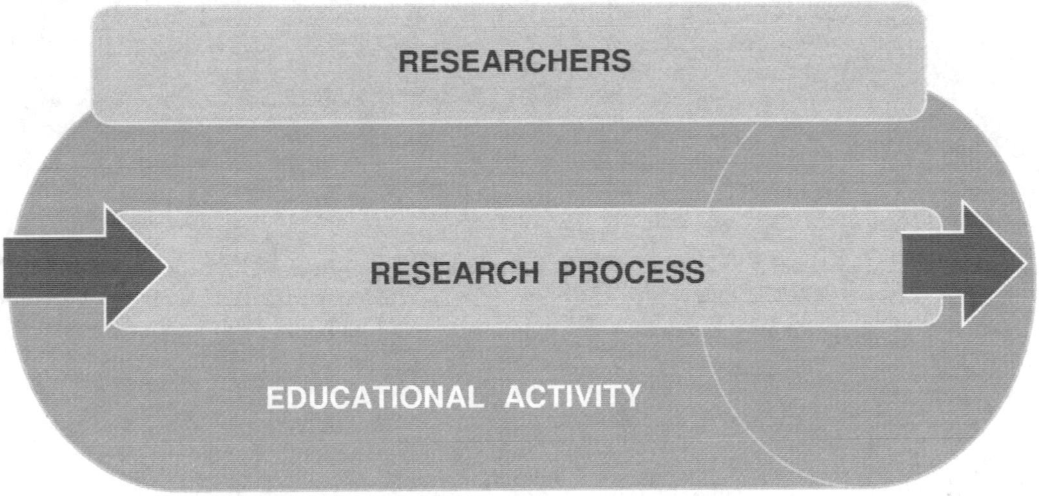

Figure 1. Scheme of University-based training.

Decaying of higher education in this respect is also noted by the management of the Russian Academy of Sciences, as lately they have observed shortage of young specialists bearing quality competencies of researchers and developers.

With regard to multiple training programs at the Universities which cover more than just engineering, digital, economic and humanistic aspects the educational programs based on interdisciplinary research are relevant today [4, 5]. For instance, this can be related to programs intended for estimation of social effects from technological innovations, studying of society via social networks using analyses of big data, enhancing of industrial safety at the production plants using digital analysis of personality.

Taking into account the aforesaid, it isn't the study of individual disciplines that educational process at the University shall begin with, but rather crossdisciplinary problematization which definitely requires development of a certain model.

4. Model of University-based education

Development of a model of the University educational process based on research shall rest upon understanding of results intended to be reached. Such a result should be the competencies of the graduates. In their studies [6-11] the authors tried to write such a competence-based model for training of engineers apt for the present-day iron and steel-making industry, as the industrial business of today is searching for certain competencies (skills and abilities to do).

The general concept of a model for the University-based training of engineers is shown in Figure 2.

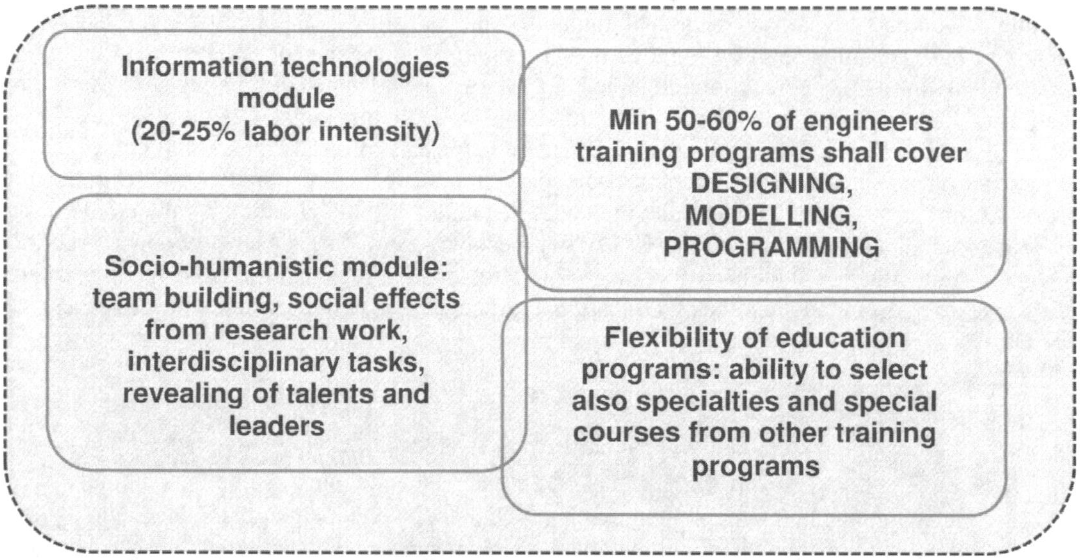

Figure 2. Components of the University-based educational program for engineers.

The very essence of all the engineering training programs shall be aimed at development of competencies in the *sphere of designing, modeling of objects and systems*, as well as description of such processes by programming languages, for which purpose it is necessary to count for a module of courses related to information technologies including data analyses and processing, skills on *CAD*- and *CAE*-systems operation.

The socio-humanistic module is required for analyses of research problematics in the aspects of history and philosophy, identifying of competencies for estimation of the social effect rendered by research and innovations, as well as activities on identifying the talents, leadership qualities and team building necessary for project activities implementation.

Training programs shall be flexible so that to enable the students to select among the courses at various levels: general University level, at the faculty, at the subdepartment. Realization of this module secures that a student can by himself map out his path in education, get certain competencies in his professional specialization, get acquainted with outstanding professors and prominent projects of the University, solve engineering and research tasks at the interdisciplinary level.

5. Conclusion and Summary

Now therefore notwithstanding the opinion of multiple experts on decaying of the higher school in Russia and its redesigning in the view of the fourth global industrial revolution (industry 4.0), we still understand that it is necessary to retain the best practices, education formats and develop engineering education which can enable training of new specialists at the level able to meet requirements of the leading industrial companies.

As for the Universities, it is obvious that in order to preserve the Universities in the light of development of distance *on-line* education formats, it will take researchers to teach but not teachers to get into science. People at the Universities must generate new knowledge, still the transit able to balance creation of new knowledge and their adoption for technology is quite relevant.

To render its maximum functionality the Universities shall be based on three main aspects: students, researchers and infrastructure (Figure 3).

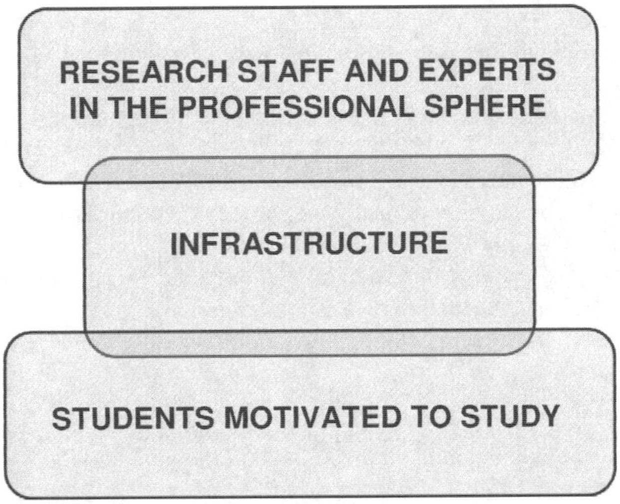

Figure 3. Conditions for functioning of the present day University.

In the process of research activities at the University innovations shall not be separated between the engineering sciences and the humanities, as all the modern science and its results rest on the junction of various fields which enables evolving of new fields in science, making the science change.

Research agenda is crucially important at the University since it allows for making a step to those who may «influence» the processes in the region, in the country, in the world (Figure 4).

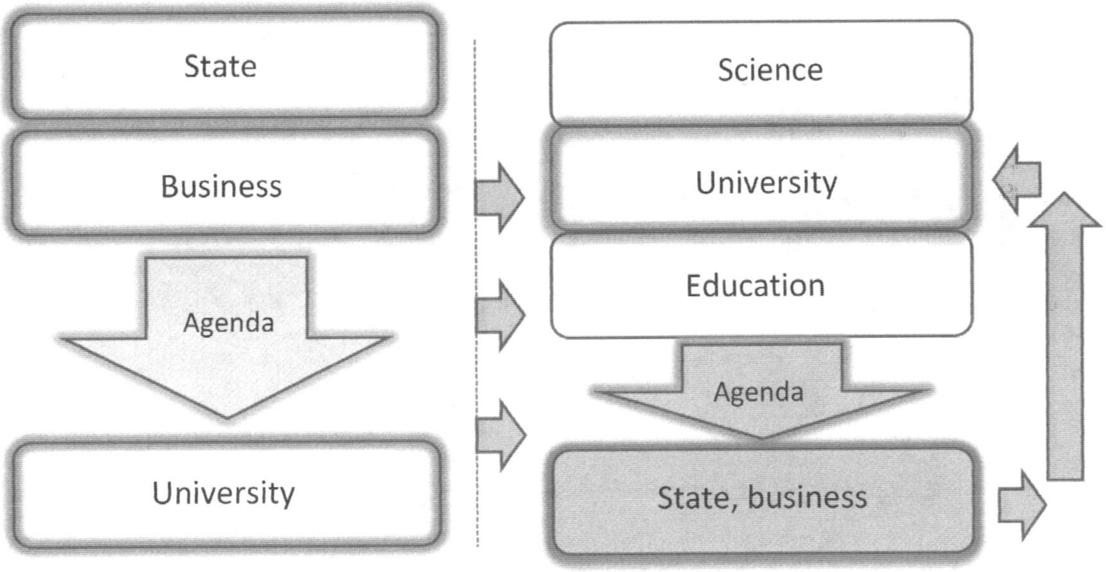

Figure 4. Reforming of University position.

References

[1] Luksha P *Education 20.35. Human-being* vol 7 (Ekaterinburg: Izdatelskie resheniya) p 152

[2] Dong B, Zheng Q, Yang J, Li H and Qiao M 2009 *Proc. The Ninth IEEE Int. Conf. on Advanced Learning Technologies* (Latvia: Riga) 125–127

[3] Uden L, Wangsa I T and Damiani E 2007 *Proc. Int. Conf. on Digital Ecosystems and Technologies (IEEE DEST 2007)* 113–117

[4] Kozhevnikov A V 2018 *AD ALTA: Journal of interdisciplinary research* **8** (1) 268–271

[5] Kozhevnikov A V 2019 *Proc. X All-Russian practical research Conf. «Information and pedagogy technologies in the modern education institution»* (Cherepovets: Cherepovets State University) 203–208

[6] Kozhevnikov A, Kozhevnikova I, Bolobanova N and Shalaevskii D 2018 *Metalurgija* **57**(4) 337–340

[7] Kozhevnikov A, Bolobanova N and Kozhevnikova I 2018 *Metalurgija* **57**(4) 367–370

[8] Kozhevnikov A, Bolobanova N, Kozhevnikova I and Shalaevskii D **2018** *AD ALTA: Journal of interdisciplinary research* **8**(2) 230 –232.

[9] Kozhevnikov A 2017 *IOP Conf. Series: Materials Science and Engineering* 287 012008 doi:10.1088/1757-899X/287/1/012008

[10] Kozhevnikov A V 2018 *Proc. IX All-Russian practical research Conf. «Information and pedagogy technologies in the modern education institution»* (Cherepovets: Cherepovets State University) 80–82

[11] Kozhevnikov A 2019 *AD ALTA: Journal of interdisciplinary research* **9**(1) 360–362

IOP Conf. Series: Materials Science and Engineering **718** (2020) 012007 doi:10.1088/1757-899X/718/1/012007

The Development and Application of Methodologies for the Design of Technological Modes of Cold Rolling

A V Kozhevnikov[1]
[1] Cherepovets State University, 5 Lunacharsky street, Cherepovets, 162600, Russia

E-mail: avkozhevnikov@chsu.ru

Abstract. The article describes the methodology of designing technological modes of cold rolling of thin steel strips on continuous mills, according to which the design process is presented in the form of seven interconnected modules that perform certain functions. As a result of the implementation of the principles, effective deformation modes are obtained, which provide a reduction in total energy costs by 5-8 % in compliance with the requirements for the quality indicators of the strip.

The general tendency to improve the quality of products affects all industries and is expressed in the desire of consumers of rolled metal to receive cold rolled steel sheet with high technological and service properties. In turn, this requires metal manufacturers to constantly introduce technologies that will improve the efficiency of existing plants and the competitiveness of cold rolled products. At the same time, there is a need to accelerate the implementation of developments, which necessitates the creation and improvement of on-line design methods for effective rolling technologies in order to improve product quality, resource and energy saving.

An analysis of the current state of production of cold-rolled steel indicates that in addition to energy saving, a key problem is ensuring the stability of consumer properties of sheet metal, which is achieved only by strictly regulating the deformation modes of rolling and controlling the technological process at the mill in specified ranges of changes in its parameters. The implementation of the above mentioned presents certain difficulties due to the inconstancy and heterogeneity of the chemical composition and structure of the steel, causing fluctuations in the value of the initial yield strength, instability of the strip thickness at the input of the mill and rolling conditions in a continuous mill, when, at the same time, the thickness and tension of the strip are adjusted, the speed and reductions values are not stable and, accordingly, power parameters of the rolling process, as well as indicators of accuracy, flatness and mechanical properties of the strips being rolled. Therefore, when designing the technological mode, it is necessary to take into account existing fluctuations in the technological and power parameters, which will allow us to develop a reliable energy-efficient cold rolling technology that provides products with a given level and stability of its quality indicators, with predicted final properties.

In the majority of works devoted to the study and design of technological modes of cold rolling, a point estimation of the parameters of the process and rolled metal without stochasticity is performed [1-6]. In [7], a methodology was proposed for calculating the instability of the parameters of the cold rolling process, taking into account the influence of random factors according to the criterion of equal forces across all mill stands, however, as practice shows, it is extremely difficult to implement. In [8-12], the application and advantages of modeling the cold rolling process taking into account the unsteadiness and stochasticity of its parameters are shown, it is concluded that it is advisable to use a dynamic model of the cold rolling process when creating new and improving existing technologies for the production of thin wide cold-rolled strips. This article is aimed at further development of modeling methods for energy-efficient rolling technologies taking into account stochastic parameters.

When developing the methodology for the design of cold rolling modes, a modular approach was applied (Figure 1). Each module implements a specific function or several interrelated functions.

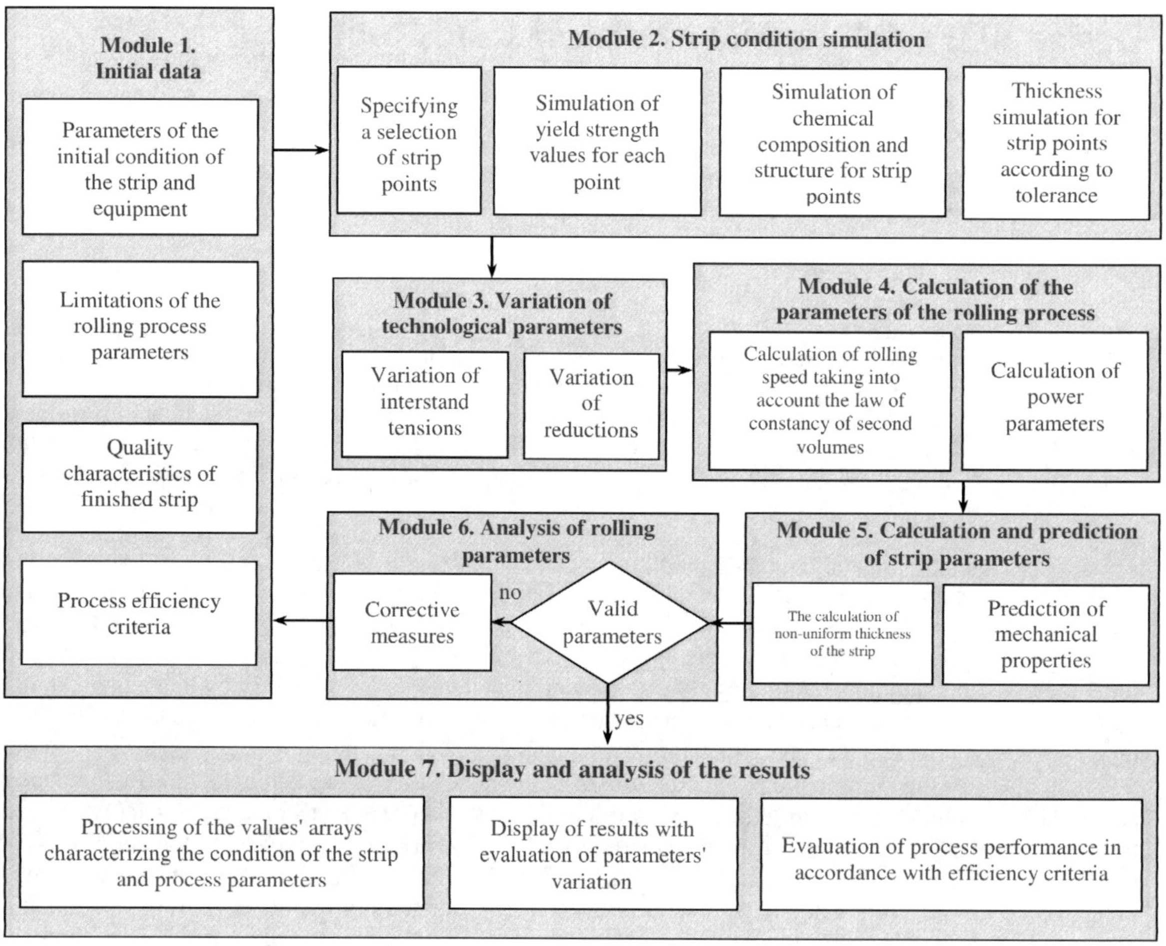

Figure 1. Methodology for the design of technological modes of cold rolling with variation of process and strip parameters.

Module 1 includes a description of the initial data for the rolled strip and allows you to set the characteristics and profile sizes of the semi-rolled and finished strip, the parameters of the mill equipment, the ranges of technological parameters such as interstand tension, reduction, rolling speeds, ranges of strip condition parameters, as well as power and quality parameters of the finished products. At the same time, the existing experience of formalized recommendations on choosing the initial values of the parameters characterizing the rolling mode can be implemented. Module 2 simulates the condition of the strip, taking into account the non-uniformity of the yield strength, chemical composition and structure, thickness along the length of the strip based on the use of the random number generation procedure according to the law of normal distribution. In module 3, it is possible to sort through all possible variations of the reduction distribution over the stands of the continuous mill and the interstand

IOP Conf. Series: Materials Science and Engineering **718** (2020) 012007 doi:10.1088/1757-899X/718/1/012007

tension according to the restrictions specified in module 1. In modules 4 and 5, for each point of the strip, the interaction parameters of the stand and strip are calculated and the condition of constancy of the second volumes of metal is checked. As a result, arrays of values are formed, which are random numbers in connection with the simulation of perturbations implemented in modules 2 and 3. After calculating the process and strip parameters in module 6, the energy and kinematic characteristics of the rolling process are checked and the quality indicators of the finished strip are evaluated. In the case of a positive check, in module 7, the processing of arrays of values characterizing the condition of the strip and process parameters is performed to select a process mode option that meets the efficiency criteria specified in module 1. If for some parameter non-compliance with the established requirements is observed, then in accordance with the design methodology, appropriate corrective measures are taken in module 1, and the calculation is repeated.

The implementation of the methodology for the design of technological modes that provide energy savings for the engines of the main drive of the mill was implemented by rolling strips 1250 mm wide and 0.5, 0.7 and 1 mm thick made of CHES30 steel at the 5-stand mill 1700 of PAO Severstal. The results obtained were evaluated and compared with the modes of strip rolling recorded at the production site. The strip condition was simulated with a standard deviation of 10% for yield strength, 0.012% for carbon content, 0.05% for manganese, 0.001 mm for grain diameter and 2% for thickness.

Table 1 shows the reduction and tension modes when rolling strips of different thicknesses. Each mode is presented in two variants: Variant 1 - actual values taken from an automatic system for measuring and controlling data during rolling; Variant 2 - calculated values in accordance with the design methodology. In all cases, the initial yield strength was 260 MPa, the rolling speed was 21 m/s, the specific tension before the first stand was 28 MPa, after the 5th stand it was 37 MPa. The thickness of the semi-finished rolled stock was 2.1 mm for strips with a thickness of 0.5 mm; 0.7 mm and 2.7 mm for strips with a thickness of 1 mm.

Table 1. Modes of reduction and specific interstand tension
when rolling strips of CHES30 steel with a width of 1250 mm on a 5-stand mill.

Variant No.	Percentage reduction in stands, %					Specific tensions in the interstand gaps, MPa			
	№1	№2	№3	№4	№5	№ 1	№ 2	№ 3	№ 4
Mode No.1 Strip thickness $h = 0.5$ mm									
1	30	30	30	30	5	145	160	150	165
2	37	26	26	32	2	132	148	186	201
Mode No.2 Strip thickness $h = 0.7$ mm									
1	24	25	24	22	3	138	152	153	154
2	32	20	20	22	5	140	165	189	200
Mode No.3 Strip thickness $h = 1$ mm									
1	22	23	22	21	4	136	144	155	169
2	28	19	19	21	4	150	164	190	200

It can be seen from the table that in the second variant, the redistribution of the reductions between the stands with the maximum load of the first stand and the unloading of the second and third stands of the mill with a sequential increase in the specific tension from the first interstand gap to the fourth one is performed. Comparison of the values of the specific energy consumption for the mill stands (Figure 2) showed that the developed modes provide 5-8% energy savings.

Experimental industrial testing of the developed modes for rolling strips of CHES30 steel at the 5-stand mill 1700 of PAO Severstal showed that:

– the average actual power saving during rolling according to the designed modes was 6.5%;

– the distribution of the strip thickness was characterized by an average value $\overline{X} = 0{,}698$ mm and a standard deviation of $S = 0{,}011$ mm, for the distribution of the yield strength, these parameters were

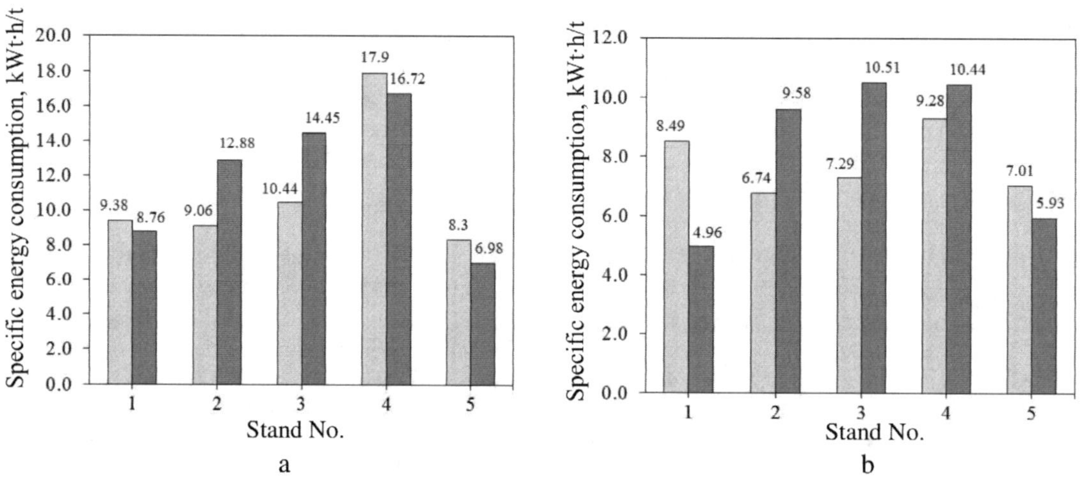

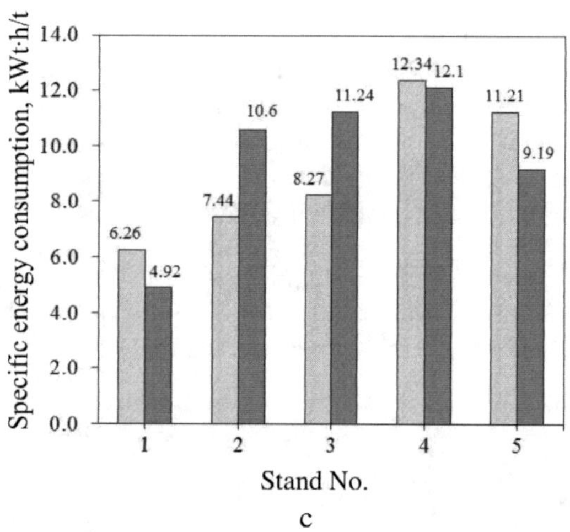

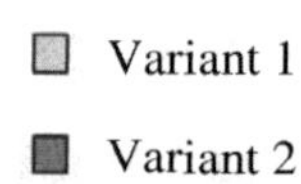

Figure 2. Comparison of specific energy costs when rolling strips of the following thickness (per stand): (a) – 0.5 mm; (b) – 0.7 mm; (c) – 1 mm.

$\overline{X}$ = 668 MPa, S = 2,17 MPa, i.e. parameters of finished strips satisfy the requirements of rolled products for cold stamping.

Conclusions

1. The methodology of developing technological modes for cold rolling of light strips in continuous mills is presented, according to which the developing process is presented in the form of seven interconnected modules that perform certain functions.

2. The implementation of these principles allows us to solve various problems in the field of the development of effective technologies for the production of thin steel strips, including the development of deformation modes that ensure a reduction in total energy costs while meeting the requirements of rolling quality indicators.

References

[1] Vasilev Ya D 1995 *Engineering models and algorithms for calculating cold rolling parameters* (Moscow: Metallurgy) p 368

[2] Garber E A and Kozhevnikova I A 2015 *Theory of rolling: a textbook for universities* (Stary Oskol: Tonkie Naukoyomkie Tekhnologii (TNT)) p 312

[3] Khimich G L and Tsalyuk M B 1973 *Optimization of cold rolling modes on a digital computer* (Moscow: Metallurgy) p 256
[4] Tselikov A I, Nikitin G S and Rokotyan S E 1980 *Theory of longitudinal rolling* (Moscow: Metallurgy) p 320
[5] Dolmatov A P, Skorokhodov V N, Nastich V P and Cheglov A E 2000 *Computer-aided design and implementation of cold rolling technology for electrical steel* (Moscow: Science and technologies) p 448
[6] Mazur V L and Nogovitsyn A V 2010 *Theory and technology of sheet rolling (numerical analysis and technical appendices)* (Dnepropetrovsk: Dnipro-VAL) p 493
[7] Rumyantsev M I, Shubin I G, Mitasov B C, Gorbunov A V and Nasonov V V 2011 *Theory and technology of metallurgical production* 208–213
[8] Kozhevnikov A, Kozhevnikova I and Bolobanova N 2018 *Journal of Chemical Technology and Metallurgy* **53**(2) 366–372
[9] Kozhevnikov A V, Kozhevnikova I A and Bolobanova N L 2017 *Metallurgist* **61**(7-8) 519–522
[10] Kozhevnikov A, Kozhevnikova I and Bolobanova N 2018 *Metalurgija* **57**(1-2) 99–102
[11] Kozhevnikov A, Kozhevnikova I, Bolobanova N and Shalaevskii D 2018 *Metalurgija* **57**(4) 337–340
[12] Kozhevnikov A V, Kozhevnikova I A and Bolobanova N L 2017 *Steel* **8** 24–29

Acknowledgements. This research was financially supported by the Ministry of Education and Science of Russian Federation within the framework of the project part of state task of Cherepovets State University (№ 11.3943.2017/4.6).

IOP Conf. Series: Materials Science and Engineering **718** (2020) 012008 doi:10.1088/1757-899X/718/1/012008

Development and Industrial Testing of Advanced Rolling Conditions at 4-Stand Mill 2100 of PAO Severstal

I A Kozhevnikova[1], N L Bolobanova[1], P V Antonov[2], S V Zhilenko[2] and A V Kozhevnikov[1]

[1] Cherepovets State University, 5 Lunacharsky street, Cherepovets, 162600, Russia
[2] PAO Severstal, 30 Mira street, Cherepovets, 162608, Russia

E-mail: iakozhevnikova@chsu.ru

Abstract. The paper presents the results of development of improved cold rolling process conditions at the 4-stand mill 2100, which allows to increase the efficiency of the process by enhancing the quality of rolled products and minimizing the occurrence of the surface defects of work rolls. Based on industrial tests, the developed advanced rolling conditions are accepted for use in production.

The increase of efficiency of the sheet rolling process by improving the quality attributes of rolled metal, increase the tool-life and reduce energy requirements is a very important aspect of competitiveness of metal produced in Russia. These improvements require technological advancement based on overall assessment of the existing production parameters, including mathematical modelling.

One of the quality indicators of cold-rolled steel that determines the condition of good application of metal and organic coating onto a strip is reduced contamination of the surface. Cold-rolled metal produced at the 4-stand mill 2100 of PAO Severstal after the reconstruction done in the end of 2016 had significant surface contamination. Contamination per a unit of sheet surface area was 25 % higher than for the metal produced at the 5-stand mill. Besides, the main equipment – work rolls of the mill – were accompanied by frequent damage of roll bodies and consequently significant roll consumption. Roll consumption at the 4-stand mill was 30 % higher compared to the 5-stand cold-rolling mill. All of that predetermined the necessity of the assessment of production technology of cold-rolled metal at the 4-stand mill 2100, including finding technology parameters which influence the degree of rolled metal contamination and the factors that determine the emergence of surface layer defects of the work rolls.

The studies of the influence of cold rolling modes on the strip surface cleanliness were based on its dependence on the positioning of neutral planes in deformation zones of the working stands, the hypothesis being the following [1, 2]: the enlargement of backward slip zone in a deformation zone helps cleaning the strip surface, because the circumferential stress in this zone along the rolling direction helps to remove the contamination from a deformation zone.

Rolling modes of more than 280 strips were taken for the study, fixed by an automatic measurement and a data control system, with the link to the quantity of grease and mechanical impurities and the reflectivity of light. For the specified modes with the application of a mathematical model, the geometric and power characteristics of deformation zones of the working stands were determined, including indicator X_i, which characterizes the position of the neutral plane *NN* in *i*-th stand (Figure 1):

$$X_i = \frac{x_{ba}}{x_{pl}}, \tag{1}$$

where x_{ba} is the length of backward slip zone of a plastic area; x_{pl} is the full length of a plastic area.

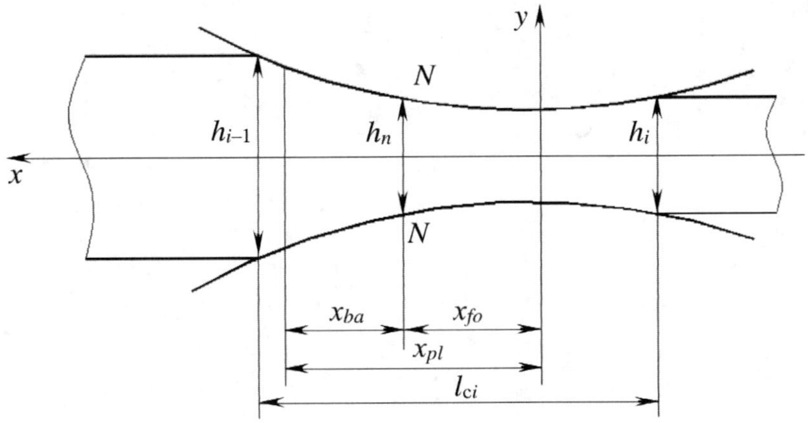

Figure 1. Deformation zone diagram: h_{i-1}, h_i is the strip thickness at the entry and exit from the i-th stand; h_n is the strip thickness in a neutral plane; x_{fo} is the plastic area forward slip zone length; l_{ci} is the deformation zone length of i-th stand; x is the rolling axis; y is the axis which connects the centres of the rolls.

The analysis of the indicators of the mechanical and grease impurities of the cold-rolled strips surface, the parameters of speed modes, reduction and tension modes, rolling forces and torques allowed us to determine that the most meaningful factors which influence contamination are: frequent percentage reduction in stand 1 (ε_1, %) and the indicators that characterize the positions of neutral planes in stand 1 (X_1) and in stand 3 (X_3). To increase the surface cleanliness of the cold-rolled strip it is necessary to decrease the reduction percentage ε_1 and indicator X_1 and increase the indicator X_3 at the same time.

In the papers dedicated to the matters of roll strength improvement [3-5], the results of the studies of technological modes' influence on roll damage and on the emergency failures of rolls are narrowed down to the recommendations on minimizing the rolling conditions, that is the reduction of rolling force in the stands, exclusion of surface and internal defects of hot-rolled metal with meeting the requirements for limit deviation of thickness from the nominal value. Furthermore, the improvement of rolling modes with the assessment of work rolls operation parameters and the important direction is the increase of rolling process efficiency.

The analysis of technological rolling modes at the 4-stand mill 2100 of the strips of steels of different hardness categories showed that:

1) Interstand specific tensions σ_1, σ_2 and σ_3 in the respective interstand spaces No. 1, No. 2 and No. 3 are basically the same, and in several cases the tension in the second interstand space σ_2 is lower than in the first one σ_1 (Figure 2). According to [2] specific tensions in every following interstand space should increase in connection to work hardening.

2) Rolling process in stand 1 is always done with front upthrust, which is showed by the graph of changes in second volume of metal, rolled in stands No. 1 and No. 2 (Figure 3, the curve is in the 1st quadrant of the coordinate plane), which leads to the significant increase of rolling force [6], and to frequent slipping of the rolls. To exclude these negative occurrences, it is needed to decrease the particular percentage reduction in the 1st stand.

3) High percentage of damages of work roll bodies of stand No. 1 is connected to frequent damage at strip gripping, because the gripping condition is not fulfilled (tg$\alpha < \mu$, where α is the gripping angle; μ is the friction coefficient). For the stand No. 1: tg$\alpha = 0.058 > \mu = 0.038$, which confirms the necessity to decrease the particular reduction in the stand.

The modelling of the technological modes of cold rolled metal production at the 4-stand mill 2100 with the application of the integrated model of cold rolling process [7, 8], which reflects the dynamic

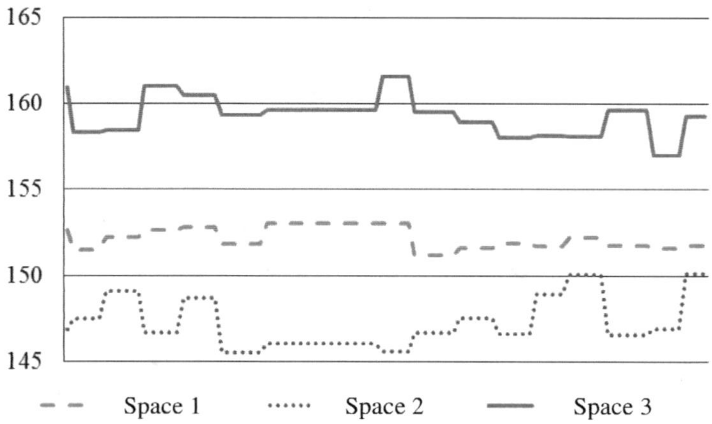

Figure 2. Distribution of specific tensions over the interstand spaces, MPa.

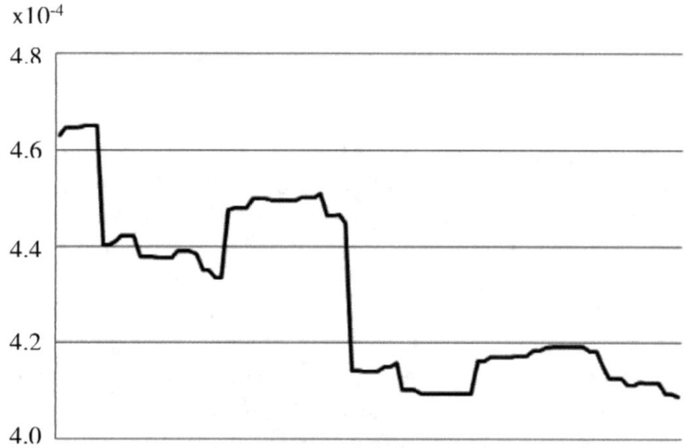

Figure 3. Changes in the difference of second volumes of metal in stands No. 1 and No. 2, m^3/s.

interconnection of the rotating system of the driveline, stand system and deformation zone, allowed us to phrase the main terms of the mill rolling concept which is the following:

1) *Deformation mode* should be assigned with maximum percentage reduction in the stand No. 2, determined from the condition of uniform loading of the first two stands by the rolling force ($P_1 \approx P_2$). In the stand No. 4 in order to get cold rolled strips with specified surface quality values and measurement precision the reduction should be in the range of 5-7 %.

2) *Tensions mode*, depending on the purpose of rolling may be set according to two alternative schemes:

– scheme 1: $\sigma_1 < \sigma_2 > \sigma_3$ or $\sigma_1/\sigma_{a1} = \sigma_2/\sigma_{a2} > \sigma_3/\sigma_{a3}$;

– scheme 2: $\sigma_1 < \sigma_2 < \sigma_3$ or $\sigma_1/\sigma_{a1} = \sigma_2/\sigma_{a2} < \sigma_3/\sigma_{a3}$,

where σ_{a1}, σ_{a2}, σ_{a3} are the actual points of strip material yield respectively at the exit from stands No. 1, No. 2 and No. 3. For the purpose of excluding breakages and tightening of cold rolled strips the specific tensions for the established process should not exceed 27 % from the existing value of the strip material yield point.

The introduced reduction and tension modes are significantly different from the ones established at the mill right now, but their implementation will allow:

– Prevention of slipping in stand No. 1 and damage of the roll bodies by decreasing the percentage reduction in the stand and fulfilling condition $\sigma_1 < \sigma_2$. For example, for the traditional steel grades 08пс

(08ps) and 08Ю (08Yu) σ_2 exceeds σ_1 for approximately 20 MPa, earlier this condition was not fulfilled (Figure 2).

– Increased cleanliness of the strip surface by decreasing the percentage reduction in stand No. 1 and assigning tensions according to scheme 1, and for stand No. 3 the inequality $\sigma_2 > \sigma_3$ is typical, which leads to the decrease of parameter X_1 (offset of the neutral plane to the side of metal entering in the deformation zone of stand No. 1) and to the increase of X_3 (offset of neutral plane to the side of metal exiting from the deformation zone of stand No. 3). According to scheme 1 the distribution of interstand tensions for steel grades 08пс (08ps) and 08Ю (08Yu) σ_2 exceeds σ_3 for approximately 25 MPa. Earlier such scheme was not used in the technological modes.

– Decreased rolling forces in the stands, and damage to the roll bodies and reduced total energy requirements by implementing reduction mode and tensions scheme 2 with gradual increase of specific tensions from the first interstand space to the last one. For example, for steels 08пс (08ps) and 08Ю (08Yu) the difference $(\sigma_2 - \sigma_1) = (\sigma_3 - \sigma_2) \approx 20$ MPa, in basic mode technology difference $(\sigma_3 - \sigma_2)$ did not exceed 10 MPa (Figure 2).

Based on the described concept of cold rolling process strip production modes were developed for the most high-demand section sizes and steel grades. As an example Tables 1 and 2 show technological and energy and power parameters of basic and advanced strip rolling modes for steel 08пс (08ps) of 1254 mm width and 0.45 mm thickness made out of material of 1.9 mm thickness at the speed of 15 m/s.

Table 1. The parameters of actual strip rolling mode dimensions 1.9→0.45×1254 mm, steel grade 08пс (08ps).

Stand No.	h_i, mm	ε_i, %	σ_i, MPa	P_i, MN	N_{engi}, kW	X_i
1	1.158	40.7	158	11.56	2015	0.8007
2	0.687	40.7	158	9.54	2082	0.9464
3	0.459	33.2	168	9.83	2106	0.8393
4	0.45	1.96	40.3	6.27	128	0.8791

Table 2. The parameters of the advanced strip rolling mode dimensions 1.9→0.45×1254 mm, steel grade 08пс (08ps).

Stand No.	h_i, mm	ε_i, %	σ_i^1, MPa	σ_i^2, MPa	P_i^1, MN	P_i^2, MN	N_{engi}^1, kW	N_{engi}^2, kW	X_i^1	X_i^2
1	1.159	39	160	150	10.2	10.3	1867	1882	0.7634	0.7775
2	0.672	42	180	170	9.5	9.8	2164	2221	0.9348	0.9303
3	0.474	29.5	155	190	8.2	8.0	1713	1643	0.8631	0.8133
4	0.45	5	52.5	52.5	9.3	9.0	398	336	0.8112	0.8369

Note. Indices 1, 2 – the value of the parameter with the existing interstand tensions distribution scheme; N_{engi} – engine capacity of the i-th working stand.

It is seen in the tables that for the advanced modes, in comparison to the basic ones, force in stand No. 1 is decreased by 11 % and the total capacity of engines, as a minimum, is decreased by 3 %. The distribution of tensions according to scheme 2 leads to additional energy efficiency which is 1 % while rolling strips less than 1.2 mm thick, and 4 % for strips with the thickness of more than 1.2 mm. The decrease of the attribute which characterizes the position of the neutral plane in stand No. 1 and its increase in stand No. 3, provided by the distribution of interstand tensions in accordance with scheme 1 helps to improve the cleanliness of the strip surface.

Conclusions and results

To check the efficiency of the developed modes, testing was done at the 4-stand mill 2100, consisting of two stages. During the first stage the simulation of coil rolling using the existing mill software; it showed the possibility to implement the proposed concept. During the second stage industrial testing was done at 34 heats, meant either for annealing or for galvanizing.

Rolling with the advanced modes provided the decrease in the total amount of mechanical impurities at the strip surface from 215 mg/m^2 to 150 mg/m^2, that is 30 %; the increase of the average rolling speed for 35 % (for the strips of up to 1.0 mm thickness the speed was 17 m/s, for over 1.0 mm it was 15 m/s). During the testing there were no damages or delamination of the surface of the roll bodies.

Based on the results of the industrial testing the developed rolling modes, which provide the increase of strip surface cleanliness and minimize the reasons of roll surface layer defects appearance, were implemented into production.

References

[1] Kozhevnikova I A and Garber E A 2010 *The development of thin-sheet rolling theory to increase the efficiency of broad-strip mills operation: a monograph* (Cherepovets: State Educational Institution of Higher Vocational Education CHSU) p 275

[2] Garber E A 2004 *Cold-rolling mills (theory, equipment, technology)* (Moscow: Chermetinformatsiya; Cherepovets: State Educational Institution of Higher Vocational Education CHSU) p 416

[3] Tretyakov A V, Garber E A and Davletbayev G G 1976 *Calculation and research of rolls* (Moscow: Metallurgy) p 256

[4] Ilyukovich B M 1964 *Roll life and consumption* (Moscow: Metallurgy) p 110

[5] Polukhin V P, Nikolaev V A, Tylkin M A, Shulman P T, Masol V A, Efimenko S P, Dunayevskiy V I, Valchuk G I, Belkin M Ya and Venzhega A S 1976 *Reliability and lifetime of cold rolling rolls* (Moscow: Metallurgy) p 448

[6] Kuvayev V N, Tokmakov P B, Klyuchnikov K Yu, Kiselev P A, Kirilenko O M, Rusalenko A V and Gordienko V V 2010 *Foundry production and metallurgy* **57**(3) 204–209

[7] Kozhevnikov A, Kozhevnikova I and Bolobanova N 2018 *Journal of Chemical Technology and Metallurgy* **53**(2) 366–372

[8] Kozhevnikov A V, Kozhevnikova I A and Bolobanova N L 2017 *Metallurgist* **61**(7-8) 519–522

[9] Kozhevnikov A, Kozhevnikova I and Bolobanova N 2018 *Metalurgija* **57**(1-2) 99–102

[10] Kozhevnikov A, Kozhevnikova I, Bolobanova N and Shalaevskii D 2018 *Metalurgija* **57**(4) 337–340

Acknowledgements. This research was financially supported by the Ministry of Education and Science of Russian Federation within the framework of the project part of state task of Cherepovets State University (№ 11.3943.2017/4.6).

IOP Conf. Series: Materials Science and Engineering **718** (2020) 012009 doi:10.1088/1757-899X/718/1/012009

The Study of the Influence of the Grain-Size Composition of Anthracite on the Agglomerating Index and the Development of Technological Recommendations for the Preparation of Fuel for Sintering for the CherMK Conditions

V Yu Kuchin[1], T V Detkova[1], A A Eliseev[1] and S V Reshiotkin[1]

[1] PAO Severstal, 30 Mira street, Cherepovets, 162608, Russia

E-mail: tvdetkova@severstal.com

Abstract. As part of the technology fine-tuning for using anthracite in sintering, Minitab 18 program was used for planning and analysis of the experiment. The deliverables of the two-level factorial experiment (2k or Design Of Experiments) were used to adjust the technology for fuel preparation in sinter production at sinter plant.

The issue of the use of the substitutes of small-sized coke for sintering is constantly in the focus of attention of scientists and production workers. Anthracite was one of the first to get into their field of view as the closest coal to small-sized coke per the mass fraction of carbon [1, 2].

In the opinion of Japanese scientists, the use of anthracite to substitute small-sized coke for sintering shifts the optimal size of the particles of solid fuel to smaller side. Upon that, it was noted that the overgrinding of anthracite can negatively impact the burning time [3].

The experience of the German researchers of the use of anthracite showed that when substituting small-sized coke with anthracite by 50 % the productivity is reduced by 7-12% for different suppliers of anthracite, but it is not reduced any more for the substitution of up to 100%. In industrial conditions when producing Thomas sinter the substitution of 20% of coke fines with anthracite did not lead to the degradation of sintering attributes. Bowl sintering for similar conditions of raw materials showed that the substitution of small-sized coke with anthracite by 50 % positively impacts the productivity, but the strength of sinter decreases [4].

When reviewing the burning process of the main kinds of sintering fuel – coke and anthracite – the Russian scientists suggested to only limit themselves with the analysis of interaction of solid carbon with the atmospheric oxygen. Based on the results of the analysis of the kinetic curve of the burnout of different kinds of fuel, it turned out that if the optimal size of the particles of coke fines is 2.0 mm, then when using a less reactive fuel – anthracite – the optimal size of its particles should be about 1.5 mm, and for a highly reactive fuel — brown-coal semicoke – the average size of the particles should be 3 mm. With these conditions, the equality of the speed of the movement of the burning front and the speed of heat transfer will be preserved [5]. The thermograms of the sintering process, made when using anthracite as solid fuel, confirm that due to low rate of burning of the particles of low-reactive anthracite the zone of high temperatures is distributed to a higher number of elementary layers of the sintered material. On one hand it leads to the decrease of maximum temperatures (in the agglomerating zone), on the other hand, due to high gas-dynamic impedance of this stretched zone, it leads to a significant

IOP Conf. Series: Materials Science and Engineering **718** (2020) 012009 doi:10.1088/1757-899X/718/1/012009

drop in the intensity of sintering process. In industrial conditions a sintering operator does not have any operative technological leverage to influence the speed of the movement of the heat wave – the "heat transfer front". There is a possibility to control the speed of movement over the layer of the burning zone, changing the size of the fuel particles, applying the fuel with different reactive capability and, lastly, by enriching with oxygen the air which gets into the layer. This way in order to increase the speed of the movement of the burning zone the main recommendation is the application of fuel with high reactive capability, decrease of the size of the fuel particles and increase of oxygen concentration in air.

The current situation (in the beginning of 2018) for the balances of coke production at CherMK set a challenge of using anthracite for sintering. The past experience of the periodical use of anthracite and also the analysis of contradicting theoretical and practical studies on the topic did not give any unambiguous advice on how to lower the technological risks when implementing this measure. For that reason, it was decided to perform a major study with the application of Minitab 18 software to plan and analyse the experiment. The laboratory sintering of agglomeration charge with the partial substitution of small-sized coke with anthracite was performed in the Centre of Raw Material Examination at the Ironmaking plant.

At the planning stage of the experiment the use of the Minitab 18 packet allowed us to pick the minimum number of necessary physical tests required for the development of the prognosis of target functions through variation of three values. The changeable parameters during the performance of a series of test sintering were the mass fractions of anthracite of 0-0.5mm, 0.5-3 mm and 3-5 mm, matching the general requirements of the preparation of solid fuel for the sintering process [1, 2, 5].

The prepared plan for the performance of test sintering with the various fractional composition of anthracite is presented in Table 1.

Table 1. The mass fraction of anthracite by the plan of experiment, % wt.

Variant number	1	2	3	4	5	6	7	8	9	10	11
-0.5 mm	16.7	0	66.7	0	33.3	16.7	0	50.0	100	50.0	33.3
+0.5 – 3.0 mm	16.7	100	16.7	0	33.3	66.7	50.0	50.0	0	0	33.3
+3.0 mm	66.7	0	16.7	100	33.3	16.7	50.0	0	0	50.0	33.3

Test sintering was performed when substituting small-sized coke with anthracite for 50%. Upon that the fractional composition of small-sized coke was maintained on the same level: 0-0.5mm – 35%, 0.5-3 mm – 55% and 3-5 mm – 10%.

The main results are presented on Figures1 and 2 as triple diagrams which illustrate the influence of the variation of the fractional composition of anthracite used as a solid substitute of small-sized coke on the sinter strength indicator per GOST 15137-77 (X_{+5}) and the specific capacity of a sintering unit by yield bin sinter (over 5 mm).

The results of test sintering (Figures 1 and 2) when using anthracite showed that the most optimal indicators of the process can be achieved in the area, where this kind of solid fuel has the bigger fraction of large-sized grains (3-5 mm).

For the application of the technology of partial substitution of small-sized coke with anthracite in industrial conditions it was necessary to make corrections in the fuel crushing technology. Figure 3 shows the current situation of the quality of fuel preparation. We can see on the diagram that the existing technology did not allow to achieve optimal characteristics for the preparation of anthracite for sintering, which led to overgrinding.

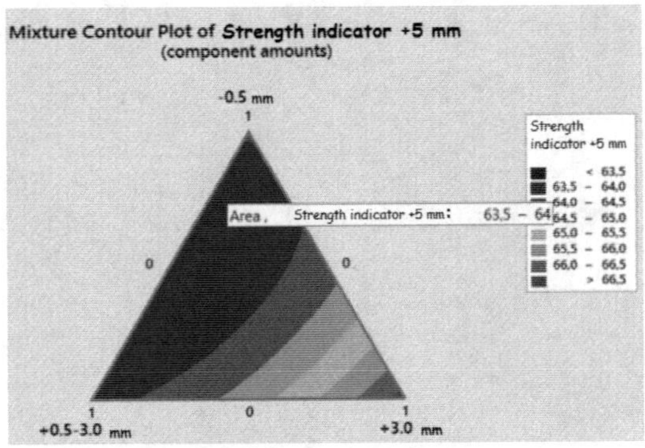

Figure 1. The analysis of the change in the strength of sinter (per GOST 15137) by variation of the fractional composition of anthracite.

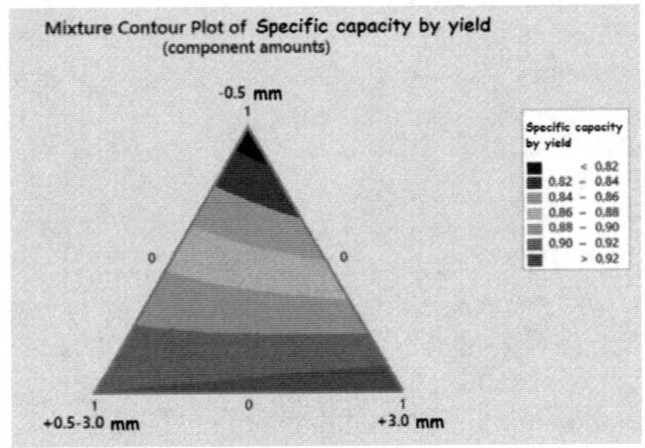

Figure 2. The analysis of the change in the specific capacity with various fractional composition of anthracite.

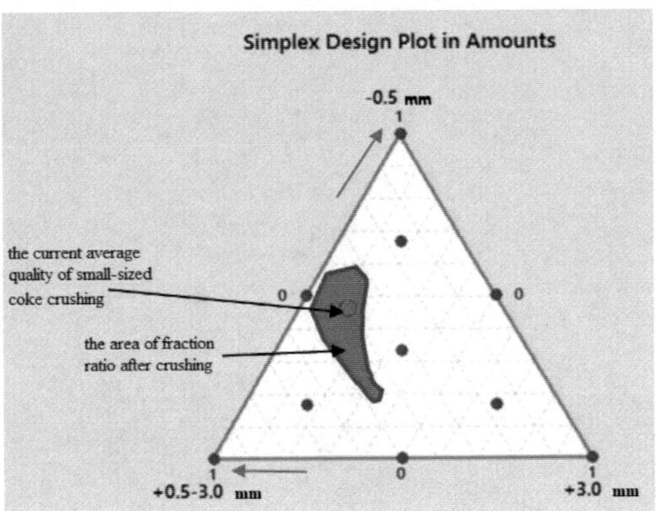

Figure 3. The area of current values of the quality of crushing of solid fuel.

The application of new parameters of crushing during the preparation of solid fuel for sintering charge showed good results in production. The quality of sinter improved significantly: fraction 0-10 mm decreased by 9.9% rel., and the number of deviations from the base per mass fraction of FeO decreased by more than 40%. The latter reveals the increase of stabilizaion of the thermal level of the process when applying a more "coarse" anthracite.

Conclusions

Thus, the use of a non-linear method of experiment planning with the application of Minitab 18 packet allowed us on tight schedule to develop the recommendations for the selection of the optimal conditions of the preparation of anthracite as a substitute of small-sized coke for sintering. The industrial application of the technology after the correction of the conditions of the preparation of solid fuel positively affected the qualitative characteristics of the blast-furnace fluxed sinter. The main effect of the conducted study certainly is the industrially applicable change of the technology of the preparation of solid fuel during the period of substitution of small-sized coke with anthracite without losing the

IOP Conf. Series: Materials Science and Engineering **718** (2020) 012009 doi:10.1088/1757-899X/718/1/012009

quality of sinter and the productivity of sinter machines to reduce the expenses for the sintering of iron-ore raw materials.

References

[1] Vegman E F, Zherebin B N, Pokhvisnev A N, Yusfin Yu S, Kurunov I F, Parenkov A E and Chernousov P I 2004 *Iron metallurgy: textbook for higher education* ed Yu S Yusfin (Moscow: IKC "Akademkniga") p 774
[2] Vegman E F 1984 *Lumping of ores and concentrates. Textbook for technical secondary education* (Moscow: Metallurgy) p 256
[3] Hosotani J and Okazaki J 2000 *The news of iron & steel industry abroad* **4** 40–41
[4] Weinlandt B and Lackman G H 1974 *Chernye metally* **22** 39–45
[5] Korotich V I, Frolov Yu A and Bezdezhskiy G N 2003 *The sintering of ore materials. Scientific publication* (Ekaterinburg: GOU VPO "UGTU-UPI") p 400

IOP Conf. Series: Materials Science and Engineering **718** (2020) 012010 doi:10.1088/1757-899X/718/1/012010

Gas Permeability Assessment of the Cohesive Zone in the Blast Furnace

A V Kuzin[1] **and N S Khlaponin**[1]

[1] Donetsk National Technical University, 58 Artem Street, Donetsk, 283001, Ukraine

E-mail: dgtu@inbox.ru

Abstract. A mathematical model is proposed for assessing the gas permeability of the cohesive zone in the blast furnace. The gas permeability is improved by increasing the intensity of the axial gas flux and introducing coke nuts in the ore layer. This is important in introducing intense pulverized-coal injection.

For the stable and highly productive operation of a blast furnace it is necessary to provide good gas permeability of the blast furnace charge. To improve gas permeability of the coke bed, the coke must undergo screening with the release of 25-40 mm fraction, rarely more than 80 mm, to achieve the optimal size of skip coke. To increase gas permeability of the lens of ore, the iron-bearing materials (mainly sinter and pellets) in several cases also are subjected to a second or a third screening to remove small fraction (less than 5 mm) [1].

Researchers have always been interested in the determination of gas permeability of the solid material layer. Between the middle of the 19th century to the beginning of 20th century when the equation for the assessment of gas permeability was being developed and clarified, the iron ore part of the blast furnace charge appeared as a dense lump iron ore, so in the process of equation development only the porosity of the layer was taken into account. During the next stage (1930-1950), when sinter and pellets started to be used in blast furnace production it became obvious that the porosity of these materials should also be considered in the equation. There are research data which mention that the determination of gas differential pressure in porous materials (for example a layer of coke or sinter) in comparison with the calculated differential as per the Darcy-Weisbach equation (where the material is taken as nonporous) gives a significant – up to 10-30 % – measure of inaccuracy [2]. This difference, by F. F. Kolesanov, is connected to the fact that the exclusion of the porosity of lumps from the calculation led to significant differences of calculated data vs. experimental data, and the materials were taken as denser (that is without taking the porosity into account) to simplify the calculation. In the beginning of the 21st century the staff of Donetsk National Technical University led by S. L. Yaroshevskiy suggested to use instead of “porosity” the indicator of “equivalent porosity” which takes into account not only the porosity of the layer, but also the porosity of the material taking into account the volume of the pores available for gas passage [3].

However, the strongest resistance to gas passage (up to 50 % from the general differential pressure in a blast furnace) is given by the zone of plastic condition – the cohesive zone. Between 1970 and 1980 under the guidance of G. Gudenau there was research done on the behaviour of the materials in the cohesive zone and the influence of the shape of the cohesive zone on the gas permeability in a blast furnace. Based on the research G. Gudenau regarded that the cohesive zone is a group of interchanging inclined layers of softened iron ore charge and coke. It was generally hypothesised that gas passes only through batch structure, and the ore layers of the softened charge are fully impermeable [4-6]. Besides, it is supposed that the cohesive zone is limited by the temperature of the beginning and the end of the softening of ore materials [7].

One of the possible ways to increase gas permeability of the lens of ore in the "dry" zone and in the cohesive zone is the integration of coke nut or coke into the iron ore layer. Under the guidance of V. I. Loginov there were extensive studies of the assessment of the influence of coke integration into the iron ore material on the blast furnace process. It was found that when mixing the entire mass of the loaded coke with the sinter compared to the normal loading system (in layers) the gas dynamic conditions of melting change and the heat-mass exchange and physical-chemical processes are activated. It is noted that when mixing the entire mass of coke with the sinter in the cohesive zone "...the areas of ore materials, which are in plastic condition, in this case will be better treated with gas and the resistance of the zone will be lower" [8].

As the laboratory studies showed, the gas permeability of the zone of softened iron ore materials improves when the coke is integrated at the temperatures typical for the cohesive zone. So, in [9] sinter of size 10-15 mm or the mix of sinter and coke was loaded into a crucible and they put the crucible into a heating oven. They imitated the load onto the materials layer by a load block at the cohesive zone layer. They injected the mix of CO, CO_2 and N_2 into the working environment, which is close to the chemical composition of gas in the cohesive zone. It is showed that in the range of temperatures 1400-1800 K the integration of coke significantly improves gas permeability of sinter in the cohesive zone: maximum decrease of gas differential pressure in sinter was observed at the integration of coke in the amount of 10 % (Figure 1).

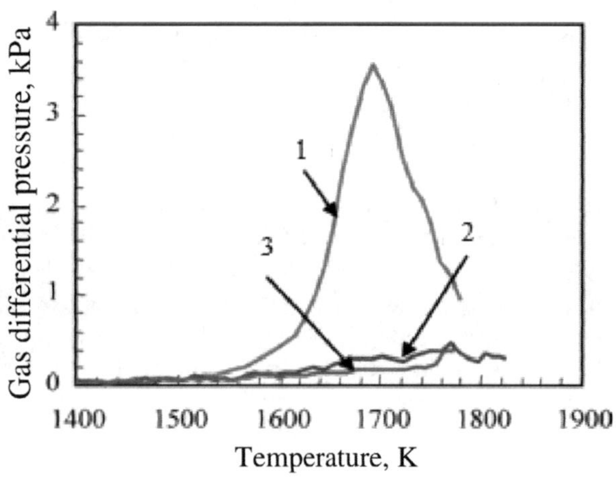

Figure 1. The change of gas differential pressure in the cohesive zone when integrating coke into sinter in the amount of 0 % (1), 5 % (2) and 10 % (3) [9].

Coming from the mentioned above a relevant objective is an analytical study of gas permeability in the cohesive zone when integrating lump fuel into the ore layer.

The objective of this study is to assess the gas permeability of the cohesive zone when integrating coke nuts into the ore layer, including when using the technology of pulverized coal injection system (PCI) into the blast furnace hearth.

Based on the literature data about the softening and melting of recovered iro n ore materials we understand that the cohesive zone is limited by the temperatures of the beginning and the end of softening. In the calculations we supposed that the cohesive zone is Λ-shaped, and the base of the cohesive zone is supported by the walls of the blast furnace belly. For the study we chose the shape of the zone Λ-shape because it provides the best gas permeability of a blast furnace especially when injecting PCI in high quantities. We took the height of the cohesive zone (H_{cz}) as equal to the height of the belly H_b to simplify the calculation (Figure 2). Over the top border of the cohesive zone there is a "dry" zone of blast furnace charge; under the bottom one is the mix of coke with melt of the ore part of the charge with overflowing slag and hot iron.

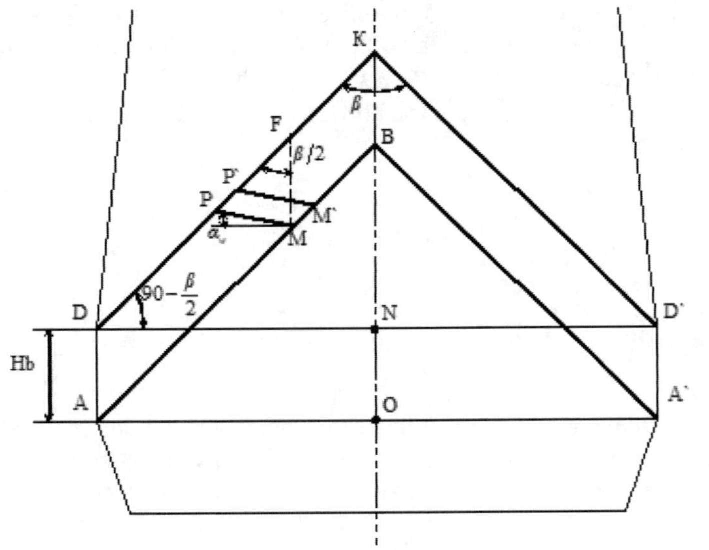

Figure 2. The scheme for the calculation of the length of gas passage through the batch structure and the heights of the excess of the cohesive zone over the border of the belly (legend in the text).

By mixing the entire mass of loaded coke with ore materials before loading into a blast furnace [8] there is reduced control over the gas passage and it complicates the process of the arrangement of mixing a high amount of coke with ore materials. That's why we think it is reasonable to mix only a part of coke or coke nut with ore materials: it preserves the possibility to control gas passage in a blast furnace while improving gas permeability of ore layers in the "dry" zone and in the cohesive zone.

The existing hypothesis formulated by G. Gudenau concerning the impermeability of the ore layer in the cohesive zone contradicts several facts. Firstly, in the "dry" zone (before the cohesive zone) the ore structures are permeable enough due to high equivalent porosity. But in the cohesive zone, in connection to the shrinkage of the iron ore layer, there is a reduct ion in the equivalent porosity of the layer. However, upon that there might be residual equivalent porosity of the ore layer. It can be permitted that through the ore layer in the cohesion zone some part of gas will pass, so the ore layer will become permeable. Secondly, at high temperatures in the cohesive zone (around 1200°C, that is at the border of the belly zone) in the lens of ore there might be a build-up of fluid melt which will partially flow to the zone located below, that is the batch structure. It will lead to some increase of porosity of the ore layer in the cohesive zone and consequently, to the increase of gas permeability of the ore structure. This assumption regarding the increase of gas permeability when a part of melt overflows was mentioned in [10]. Besides, the authors of [11] while developing a model for the assessment of gas permeability of the blast furnace charge supposed that in case of the integration of coke into the iron-bearing material in the cohesive zone the ore layer becomes permeable for gas.

Coming from that we can suppose that the ore structures are partly gas permeable and to improve the accuracy of calculation these factors must be considered in the methods of calculation of gas permeability of the cohesive zone. In our opinion, this improves the accuracy of calculation. In connection to that in this study we made an attempt to assess the gas permeability in the cohesive zone while integrating coke nut into the ore part of the charge while increasing the consumption of PCI with the regard to the assumed gas permeability of the ore structure.

As per the scheme (Figure 2) the height of the excess of the cohesive zone over the border of the belly was determined by the formula, m

$$KN = DN \cdot \mathrm{tg}\left(90^{\circ} - \frac{\beta}{2}\right), \tag{1}$$

where β is the angle in the top of the cohesive zone in degrees.

The length of gas passage through batch structure or through the layer of ore material *PM* can be determined by geometry based on the similarity of triangles. In the triangle *PMF* the side *FM* is known (taken as equal to the height of the melt *AD*) and two angles $\angle PFM = \beta/2$; $\angle PMF = 90 - \alpha_W$ (where α_W – is the inclination angle of batch structures and ore layers to the horizon). Besides, these data allow us to determine the side *PM* by the formula, m

$$PM = FM \cdot \frac{\sin(\beta/2)}{\sin(180 - \beta/2 - (90 - \alpha_W))}. \tag{2}$$

The length of the generatrix of the cone can be computed by the formula

$$DK = \sqrt{DN^2 + KN^2}, \tag{3}$$

where *DN* is the radius of the melt in m.

The thickness of the layer of coke and ore component (or the mix of the ore component and injected coke nut) was assessed based on their quantity in the charge considering their position on the melt layer.

To determine the volume of the gas passing through the batch and ore structures it is necessary to assess the lateral area of the cohesive zone. Considering the earlier assumption regarding the shape of the cohesive zone it will have cone shape. Then the lateral area of the cohesive zone (Figure 2) taking into account formula (3) can be assessed by the formula, m^2

$$S = \pi \cdot DN \cdot DK. \tag{4}$$

We think that gas passage through batch and ore structures is distributed proportionally to the surface areas of coke S_c and ore S_o structures for gas passage and also the porosity of the coke and the ore layer in the cohesive zone. Upon that we need to consider two factors for the cohesive zone:

- the decrease of surface area for gas passage in ore structures because of the softening and partial melting of the ore layer (layer shrinkage);
- the reduction of the surface area for gas passage in coke bed due to frequent scarifying and degradation of coke.

Based on the laboratory data the shrinkage of sinter upon softening and partial melting is 22-26 % [12, 13]. By the data of [14] ore melt can flow into batch structure in the volume of up to 0.08 m^3/m^3. Considering the factors specified above we take the residual equivalent porosity of coke in the cohesive zone (the volume and, consequently, the height of the coke layer on the level of the bottom of the shaft of the furnace do not change) as equal to $\varepsilon^{c}_{res} = 0.42$ m^3/m^3.

The surface area of coke bed S_c and lens of ore S_o was assessed by the formulas, m^2

$$S_c = S \cdot \frac{h_c}{h_c + h_o \cdot (1 - \Delta V)}; \tag{5}$$

$$S_o = S \cdot \frac{h_o \cdot (1 - \Delta V)}{h_c + h_o \cdot (1 - \Delta V)}, \tag{6}$$

where h_c and h_o are the heights of coke beds and lenses of ore respectively in m; ΔV is the shrinkage of the ore layer (ore structures) in the cohesive zone as a unit fraction.

The initial equivalent porosity of the ore layer was determined considering the equivalent porosity of the sinter and pellets by the formula

$$\varepsilon_{beg}^{por} = \varepsilon_{eq}^{s} \cdot \%_{s} + \varepsilon_{eq}^{p} \cdot \%_{p}, \tag{7}$$

where ε_{eq}^{s} and ε_{eq}^{p} are the equivalent porosities of the sinter and pellets in m^3/m^3; $\%_{s}$ and $\%_{p}$ are the fractional proportions of sinter and pellets in the charge as a unit fraction.

The residual equivalent porosity of the ore layer (ore structures) without the injection of coke nut was assessed by the formula, m^3/m^3

$$\varepsilon_{res}^{por} = \frac{\varepsilon_{beg}^{por} - \Delta V}{(1-\varepsilon_{beg}^{por}) + (\varepsilon_{beg}^{por} - \Delta V)}. \tag{8}$$

The residual equivalent porosity of the ore layer considering the injection of coke nut was determined by the formula, m^3/m^3

$$\varepsilon_{res}^{sym} = \varepsilon_{res}^{por} \cdot \%_{por} + \varepsilon_{res}^{nut} \cdot \%_{nut}, \tag{9}$$

where ε_{res}^{nut} is the residual equivalent porosity of coke nut in the lens of ore in the cohesive zone in m^3/m^3; $\%_{o}$ and $\%_{nut}$ are the fractional proportions of ore materials and coke nut as unit fractions.

The fraction of the surface area of batch structure for gas passage (the fraction of gas passing through batch structure) was assessed by the formula

$$Q_c = \frac{S_c \cdot \varepsilon_{res}^{c}}{S_c \cdot \varepsilon_{res}^{c} + S_o \cdot \varepsilon_{res}^{sym}}. \tag{10}$$

Then the fraction of gas passing through the ore structure is assessed by the difference

$$Q_o = 1 - Q_c. \tag{11}$$

The quantity of the hearth gas which is formed in the bottom part of a blast furnace was determined by the formula, m^3/s

$$V_{HG} = \frac{P_{day} \cdot V_{HG}^{o}}{24 \cdot 60 \cdot 60}, \tag{12}$$

where V_{HG}^{o} is the discharge of hearth gases for 1 tonne of hot metal in nm^3/t of hot metal; P_{day} – is the daily performance of the furnace, t / day.

Then the amount of gases which pass through the cohesive zone taking into account the temperature and the pressure was assessed by the formula, m^3/s

$$V_{HG}^{t} = V_{HG} \cdot \frac{t_{av}^{CS} + 273}{273} \cdot \frac{p_0}{p_b}, \tag{13}$$

where t_{av}^{CS} is the average temperature in the cohesive zone in °C, it is taken as equal to the average value between the temperature of the end and the beginning of the softening of ore materials; p_0 and p_b are the absolute values of the atmospheric pressure and blast respectively, kPa.

The velocity of gas passage through batch structure was determined by the formula, m/s

$$U_0 = \frac{V_{HG}^{t} \cdot Q_c}{S_c \cdot \varepsilon_{res}^{c}}. \tag{14}$$

The influence of the injection of fuel additives on the coke rate was assessed by the formula, kg / t of hot metal

$$C_r \approx C_0 \cdot [1 - 0.01 \cdot (Fe - Fe_0)] - 0.9 \cdot PCI, \tag{15}$$

where C_0 is the coke rate with the initial iron content in the charge Fe_0 (%), absence of *PCI*, but with the same high temperature of blasting and other parameters of blast furnace melting in kg/t of hot metal; *PCI* is the flow rate of *PCI* in kg / t of hot metal; Fe is the percentage of iron content in the charge during a base period.

The gradient of the gas temperature was determined by the formula, K/m

$$a_T = \frac{T_{se} - T_{sb}}{H_{cz}}, \tag{16}$$

where T_{sb} is the absolute temperature of the beginning of softening in K; T_{se} is the absolute temperature of the end of softening in K; H_{cz} is the height of the cohesive zone in m.

The calculation of the absolute pressure in the cohesive zone was assessed by the formula [14], kPa

$$p = \sqrt{p_s^2 + \frac{f}{\sin \alpha_W} \cdot \frac{\left(1 - \varepsilon_{res}^c\right)}{\left(\varepsilon_{res}^c\right)^3 \cdot d_e} \cdot \frac{p_0 \cdot \rho_g \cdot U_0^2}{T_0 \cdot F} \cdot \left(T_{sb} \cdot H_{cz} + \frac{a_T \cdot H_{cz}^2}{2}\right)}, \tag{17}$$

where p_s is the gas pressure on the border of the charge and the belly in Pa; f is the specific resistance; d_e is the equivalent (harmonic mean) diameter of the lumps of coke in the cohesive zone in m; F is the factor of the shape of a lump; T_0, p_0 are the normal absolute temperature and pressure (T_0 = 273 K, p_0 =101 kPa); ρ_g is the normal density of gas, kg/m^3.

Then the differential gas pressure in the cohesive zone can be determined by the formula, kPa

$$\Delta p_{cz} = p - p_s. \tag{18}$$

When changing the conditions of blast furnace melting (putting coke nut into the ore layer, using *PCI*) the change of p_s can be assessed by the change of the differential gas pressure in the shaft of the furnace (in the “dry” zone of the charge). However, the consideration of the clarified p_s value will not significantly affect the determination of the value Δp. That’s why in our calculations the value p_s is taken as unchanged based on the actual indicators of the operation of the blast furnace.

As the initial data, we used the data of the operation of the blast furnace no. 3 of Yenakiieve Iron and Steel Works in the volume of 1719 m^3. In the calculations the absolute pressure of blasting p_b is taken as equal to 350 kPa; the pressure of gas at the border of charge and belly p_c = 300 kPa; the inclination angle of the batch structure and ore layers to the horizon α_w = 22°; the density of gas ρ_g = 1.16 kg/m^3; the equivalent (harmonic mean) diameter of coke lumps in the cohesive zone d_e = 0.05 m; the height of the cohesive zone H_{cz} = 2.0 m; the absolute temperature of the beginning of softening T_{sb} = 1273 K; the absolute temperature of the end of softening T_{es} = 1473 K; the residual porosity of the coke nut in the lens of ore in the cohesive zone ε_{res}^{nut} = 0.45 m^3/m^3; the shrinkage of the ore layer (ore structure) in the cohesive zone ΔV = 0.26; the factor of lump shape F = 0.75. In the calculation it was taken that the size of coke nut in the cohesive zone was the same as when it was loaded.

The results of the calculation by the formulas (1)-(4) are specified in Table 1. From Table 1 we can see that when reducing the angle β from 120 to 60° the cohesive zone significantly stretches out to the shaft from 2.9 to 8.8 m (more than thrice), the length of gas passage through the cohesive zone decreases from 2.2 to 1 m, the surface area of side surface of the cohesive zone increases from 94.5 to 163.4 m^2 (more than 1.7 times). This circumstance will help the increase of batch structure and, consequently, the improvement of gas permeability of the cohesive zone. This is confirmed by the experience of the operation of blast furnaces with low coke consumption: the increase of *PCI* consumption led to decreased gas permeability of the cohesive zone and to preserve it or to improve it we had to stretch the cohesive zone out [15]. Besides, the calculation done by the formulas (10) and (11) shows that the injection of coke nut in the volume of 0 - 30 % into the ore layer will help to increase the gas fraction passing through the ore layer from 6.9 to 36.1 %.

Table 1. The determination of the main sizes of the cohesive zone for the conditions of the blast furnace of 1719 m^3 capacity.

Indicators	The angle in the top of the cohesive zone β		
	60°	90°	120°
1. The height of the cohesive zone excess over the border of the belly, m	8.8	5.1	2.9
2. The length of the gas passage through batch (ore), m	1.0	1.5	2.2
3. The length of the generatrix of the cone, m	10.2	7.2	5.9
4. The surface area of the side surface of the cohesive zone, m^2	163.4	115.4	94.5

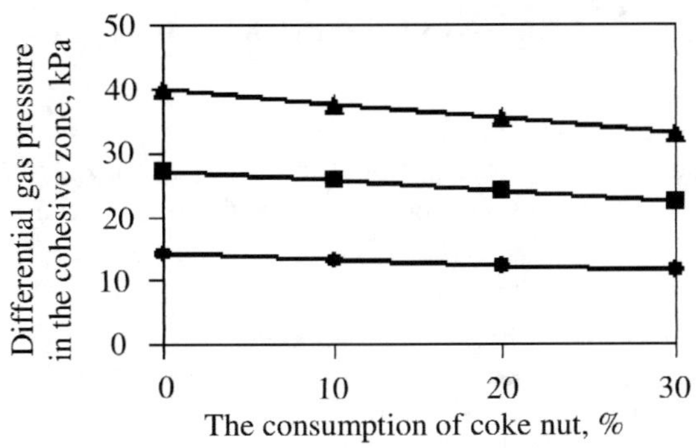

Figure 3. The influence of the injection of coke nut in the iron ore part of the charge on the differential gas pressure in the cohesive zone at low (▲), medium (■) and high (•) intensity of the axial gas flow.

A series of calculations was done to assess the gas permeability of the cohesive zone while injecting coke nut into the ore layer in the amount of up to 30 % at low (β=120°), medium (β=90°) and high (β=60°) intensity of the axial gas flow. Figure 3 shows that when increasing the fraction of coke nut from 0 to 30 %, loaded into the iron ore part of the charge, with different intensity of the axial gas flow the differential gas pressure decreases by 17.5-18.6 %. Besides, changing the intensity of axial gas flow from low to high (angle β in the top of the cohesive zone reduces from 120 to 60°) will lead to the reduction of the resistance to gas passage from 33-40 to 11.4-14 kPa (2.9 times).

Besides, a series of calculations was done to assess the gas permeability of the cohesive zone for the conditions specified above and the *PCI* in the amount up 250 kg/t of hot metal. Figure 4 (a) shows that while increasing the consumption of *PCI* from 0 to 250 kg/t of hot metal and zero consumption of coke nut the differential gas pressure in the cohesive zone increases from 40 to 55.8 kPa (or by 39.5 %). To keep the base value of the differential gas pressure while increasing the consumption of *PCI* one of the compensating measures can be the increase of the consumption of coke nut mixed with the iron ore part of the charge. So, for example, at low intensity of gas passage to keep the base value of the differential gas pressure, equal to 40 kPa, and the increase of the consumption of *PCI* up to 150 kg/t of hot metal the consumption of coke nut should be about 30 % (Figure 4 (a)). At high intensity of gas flow the differential gas pressure with the consumption of *PCI* up to 250 kg/t of hot metal will be significantly lower than the base value of the differential gas pressure (40 kPa) (Figure 4 (b)).

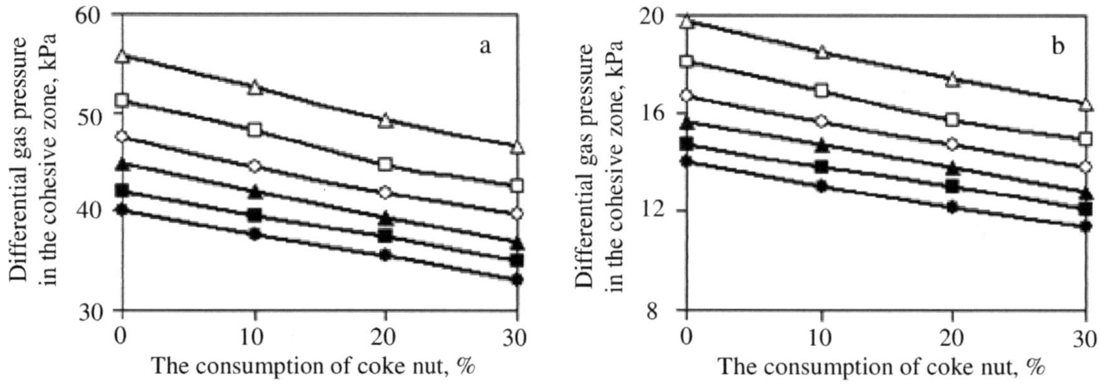

Figure 4. The influence of the injection of coke nut into the iron ore part of the charge on the differential gas pressure in the cohesive zone at low (*a*) and high (*b*) intensity of the axial gas flow with PCI, kg/t of hot metal: • – 0; ▪ – 50; ▲ – 100; ○ – 150; □ – 200; Δ – 250.

Conclusion

The calculations of gas permeability are made based on the assumption that ore structures are impermeable to gas. The analysis of the literature data and the calculations show that in a blast furnace ore structures in the cohesive zone, as a rule, are permeable to gas, which significantly influences the distribution of gas flow. Up until now the method of calculated determination of gas permeability of ore structures in the cohesive zone is not fully perfected. In this study we suggested a method of calculated determination of gas permeability of the cohesive zone, which considers the influence of the loading of coke nut into the ore layer. As a main statement we take that the ore structure in the cohesive zone is permeable for gas. Besides, the developed method of assessment allows us to determine the gas permeability in the cohesive zone when changing a range of technological parameters of blast furnace melting.

Changing the intensity of the axial gas flow in a blast furnace to a higher one will help to stretch out the height of the cohesive zone over the border of the belly and to improve the gas permeability of the cohesive zone. We assessed the influence of different intensity of the axial gas flow in the cohesive zone while injecting coke nut into the ore layer. It showed that when a blast furnace is operating with skip coke prepared by fractional composition and with the injection of coke nut into the iron ore part of the charge in the amount of up to 30 % the transfer from low intensity of axial gas flow to the high one will help to improve the gas permeability of the cohesive zone by 2.9 times. This measure is quite relevant especially in case of assimilation of blast furnace melting with the high consumption of *PCI*.

References

[1] Yaroshevskiy S L and Khlaponin N S 2006 *Proc. Int. Scientific and Technical Conf. Pulverized fuel - the alternative to natural gas for iron melting* (Donetsk: UNITEKH) p 59–64
[2] Kolesanov F F 1956 *Gas flow through the layer of lump materials.* (Moscow: Metallurgizdat) p 88
[3] Yaroshevskiy S L, Nozdrachev V A, Chebotarev A P and et al 2000 *Metallurg* **12** 32–35
[4] Tarasov V P, Khairetdinova O T and Tomash A A 2002 *Izvestiya. Ferrous metallurgy* **4** 64–66
[5] Gudenau G, Kreibich K and Nomia Y 1979 *Chernye metally* **22** 7–13
[6] Simkin A I, Ostapenko V A and Tomash A A 2003 *Vestnik Priazovskogo gosudarstvennogo tekhnicheskogo universiteta* 13 20–24
[7] Shürman E, Gudenau G and Peters K-H 1982 *Chernye metally* **6-7** 9–14

[8] Loginov V I, Musienko K A, Berin A L and et al 1984 *Metallurgical and mining industry* **3** 10–12
[9] Watakabe S, Takeda K, Nishimura H and et al 2006 *ISIJ International* 46(**4**) 513–522
[10] Makhanek N G and Onorin O P 1967 *Izvestiya vuzov. Ferrous metallurgy* **2** 29–31
[11] Matsuhashi S, Kurosawa H, Natsui S and et al 2012 *ISIJ International* 52(**11**) 1990–1999
[12] Chernavin A Yu, Nechkin G A and et al 2010 *Steel* **5** 20–23
[13] Kolesanov F F, Khlaponin N S, Krivosheev V N and Chikurov V I 1983 *The perfection of sinter process* (Kiev: Technica) p 110
[14] Tomash A A and Kuzin A V 2010 *Science and technology of Donetsk National Technical University. Seriya: Metallurgy* **12** 89–100
[15] Gudenau G, Kreibich K, Kortach B and Birkheuser L 1988 Chernye metally **19** 10–19.

Acknowledgements. The study was made under the guidance of D.Sc. in engineering (doctor of technical sciences), prof. S. L. Yaroshevskiy.

IOP Conf. Series: Materials Science and Engineering **718** (2020) 012011 doi:10.1088/1757-899X/718/1/012011

The Method of Efficiency Definition of New Functional Task in Electric Arc Furnace Control System

O S Logunova[1], V A Oshurkov[1] and V V Pavlov[2]

[1] Nosov Magnitogorsk State Technical University, 38 Lenin Street, Magnitogorsk, 455000, Russia

[2] PJSC Magnitogorsk Iron and Steel Works, 93 Kirov Street, Magnitogorsk, 455000, Russia

E-mail: logunova66@mail.ru

Abstrast. The task of the study is to create the measures and methods of efficiency definition of new functional task in electric arc furnace control system. The method of system analysis was applied. The research was carried out in the shops of Magnitogorsk Iron and Steel Works (MMK). Method of efficiency definition of new functional task module in electric arc furnace control system was developed. The capacity loading reduction of bucket quantity and materials in charging process in the electric arc furnace for any planned task was presented.

1. Introduction

Automatic control system of technological process (ACS TP) is carried out as a part of iron and steel melting facility. Automatic control system of technological process with scaled out architecture is in the priority of iron and steel melting facility in the development according to the new trend. The provision of new functional task into the control system is not complicated and expensive work for these systems [1–3]. The ACS TP functional task determining the poured bulk density properties of metal scrap fragments in melting facilities workspaces has not been researched enough yet. The problem of the study is the subject of many researches in the works of russian and foreign scientists. The authors put forward the ideas determining the intimate poured bulk density properties of metal scrap fragments of melting facilities workspaces [4,5] but with no information of the inadvertent heap in charging scrap metal. The inadvertent heap in charging scrap metal leads to increase of energy consumption, electric arc furnace effectiveness slowdown, increase the probability of breakdown electrodes. Decisions making on the number of the scoops is the task of the expert planner based on skills, method of material charging into the electric arc furnace and metal scrap net weight into the scoop. The block diagram of the decisions making on the number of the charging scoops is shown on Figure 1. The following designations are accepted on Figure 1: I_0 – information support; I_1 – information on the availability of liquid iron; I_2 – metal scrap net weight into the scoop information; I_3 – scoops identification number into sequences of the scoop; M_1 – choice the scoop from sequences of the scoop; M_2 – charging scrap metal into loading bucket and electric arc furnace; m_i – metal scrap net weight into the i-scoop.

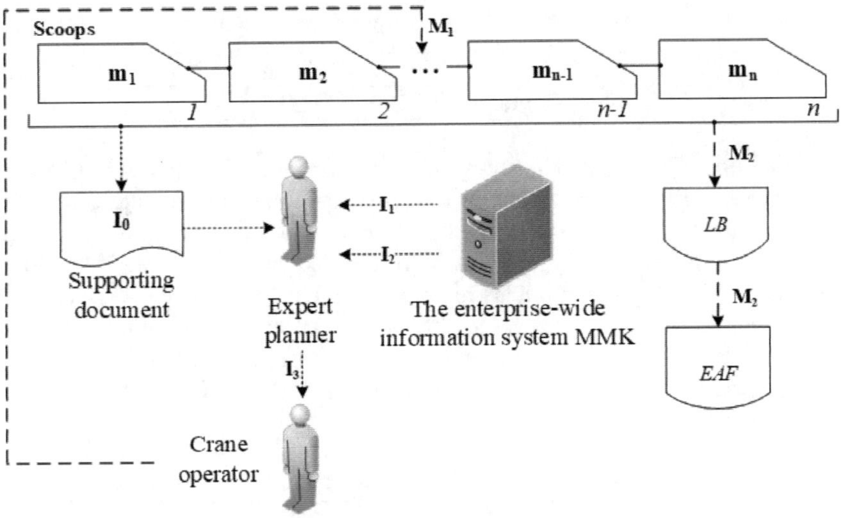

Figure 1. The block diagram of the decisions making on the number of the charging scoops.

The authors of the research put forward the idea to scale out the functional architecture of ACS TP steelmaking and provision of the new functional task – the intellectualization of the decisions making process on the charging scoops number with the inadvertent heap in charging scrap metal process. The authors of the research have proved the following:

– hardware and software conceptual schema was constructed of the ACS TP steelmaking with the new functional task;

– mathematical support was developed as the task of the technique of metal scrap fragments function construction of the poured bulk density in melting facilities workspace [6,7];

– appropriateness of results modeling experiment and simulation experiment results of the inadvertent heap in charging scrap metal was proved [8,9].

So the challenge is to estimate the efficiency of the new functional task of the electric arc furnace control system (ACS TP steelmaking) by the new method of efficiency definition of new functional task in electric arc furnace control system.

2. Experimental research

One of the indicators for estimate of steel cost is energy consumption and energy consumption. Duration process of heat is one of the factors affecting the electric-steelmaking furnace productivity. Duration process of heat depends on scrap charge duration and additional charge duration. Energy costs determine arc under current duration. The arc under current duration reliant is dependent on the number and size of metal scrap into the scoop. Thus, time index is based on the number of scoops and/or scoops available in scrap charge and can be used for the efficiency estimate of the new functional task in the electric arc furnace control system.

In the course of the research it was gathered the empirical information from the heat report for arc under current duration and heat duration time pattern recognition. The empirical information includes: indicators of the number of heats per month where one, two or three scoops and/or a scoops were used for scrap charge; the average value of the arc under current duration; the average value of the share of time duration arc under current. The number of heats types were rated to the total heats. The average values of the indicators are shown in table 1, where D_1 – the share of heats in the total quantity per month with one scoop in the scrap charge; D_2 – the share of heats in the total quantity per month with two scoops in the scrap charge; D_3 – the share of heats in the total quantity per month with three scoops in the scrap charge; T_c – heat duration time average, min; D_I – average value of the share of time duration arc under current.

Table 1. The average values of the indicators.

D_1	D_2	D_3	T_c (min)	D_I
0.19	0.68	0.12	31.96	0.53

The correlation mining results data has shown the occurrence availability of dependence of average value of the share of time duration arc under current and heat duration time indicator with two scoops in the scrap charge. The statistical analysis results of correlation data mining are shown in Figures 2 and 3, where intercept – intercept term in linear regression equation; St. Err. of b – standard errors s_j estimates of the coefficients of the linear regression equation; p-level – significance level for regression equation coefficients; b – the values of the coefficients of the linear regression equation; $t(31)$ - the value of the Student statistics to test the hypothesis about the zero value of the corresponding coefficients; R – determination coefficient; F – the value of the Fisher statistic (1 and 31 are the degree of freedom).

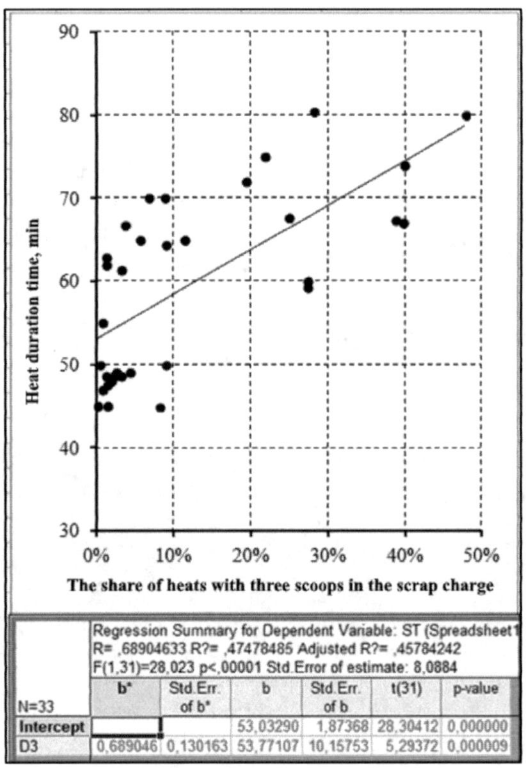

Figure 2. The statistical analysis results of correlation mining data to determine heat duration time indicator.

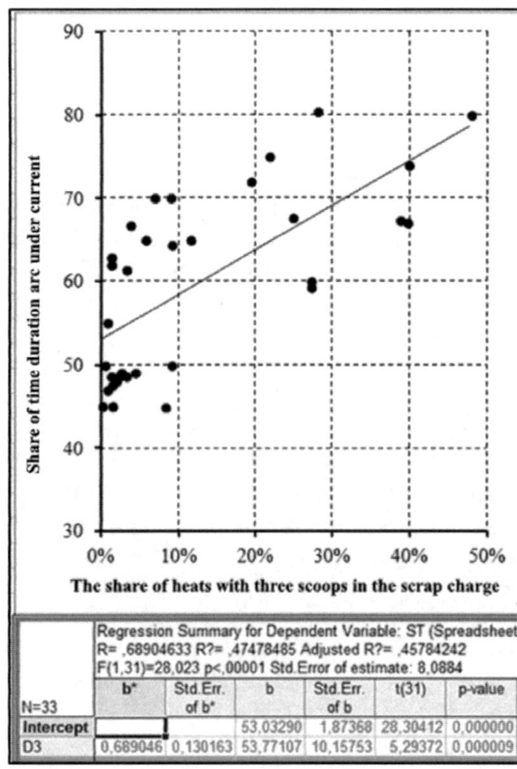

Figure 3. The statistical analysis results of correlation mining data to determine share of time duration arc under current indicator.

The statistical analysis results [10] has shown the reliable regression equations for:

– heat duration time:

$$T_c = 53.77 \cdot D_3 + 53.03; \tag{1}$$

– share of time duration arc under current:

$$D_I = 0.257 \cdot D_3 + 0.501. \tag{2}$$

IOP Conf. Series: Materials Science and Engineering **718** (2020) 012011 doi:10.1088/1757-899X/718/1/012011

The coefficient regression equations are significant at 5 %. Therefore, it is permissible to use and determine the values of these indicators from experimental data.

3. The new method of efficiency definition of new functional task in electric arc furnace control system

The empirical equation has shown the decrease of heat duration time and decrease share of time duration arc under current with two scoops in the scrap charge. The new functional task of the electric arc furnace control system permits to make a forecast occurrence the heap of charging scrap metal and the poured bulk density of metal scrap fragments distributed in space. It becomes possible to exclude scoops in charging process with suboptimal mass and suboptimal poured bulk density of metal scrap fragments. Thus, as the performance indicators for the sophisticated control system with the new functional task of the electric arc furnace control system may be:

$$I_1 = \frac{T_c}{T_{c0}} \cdot 100\% ; \qquad (3)$$

$$I_2 = \frac{D_I}{D_{I0}} \cdot 100\% , \qquad (4)$$

where T_{c0} – the predicted cycle time with a reduction in the number of heats with three scoops in the charging, min; D_{I0} – the predicted fraction of the time the arc is under current in reducing the number of heats with three scoops in the charging, min.

The aggregated performance factor of the steelmaking process control systems was formulated and is based on (3) and (4):

$$I = \alpha_1 I_1 + \alpha_2 I_2 , \qquad (5)$$

where α_1, α_2 - weighted coefficients providing the flexibility of the process control systems when choosing priorities (α_1, $\alpha_2 = 1$).

Method of efficiency definition of new functional task module in electric arc furnace control system comprises:

1) using the data from the enterprise-wide information system to perform the dependencies adaptation (1) and (2) not less than 3 years;

2) using data from an enterprise-wide information system to determine the number of heats with three scoops in the scrap charge in the preceding month;

3) using dependencies adaptation (1) and (2) to determine the values of the heat duration time and share of time duration arc under current in the previous month without the new functional task module;

4) installing the new functional task module it is necessary to perform points 2 and 3;

5) to determine the performance indicator by the formula (5);

6) to perform an efficiency assessment according to new functional task module: if I>1, where the new functional task module is efficient in other case the new functional task module is not efficient.

4. The results

The source data for the method of efficiency definition of new functional task module in electric arc furnace control system are shown in Table 2.

The results of determining the final efficiency by the formula (5) with the value of the main weight coefficient α_1 with a step of 0.1 are shown in Figure 4.

5. Conclusions

1) The empirical equations were made and with the information of the number of scoops, the heat duration time and share of time duration arc under current.

2) The performance factor of the new functional task module was built with the information about the number of scoops, the heat duration time and share of time duration arc under current.

3) The new method of efficiency definition of new functional task in electric arc furnace control systems was created using the aggregated performance factor.

Table 2. The source data for the method of efficiency definition of new functional task module in electric arc furnace control system.

The indicators	Indicators value				
The source data					
1. The share of heats with third scoop in the scrap charge in the previous month	0.07	0.1	0.3	0.4	0.5
2. The share of heats with third scoop in the scrap charge in the current month *	0.1	0.1	0.1	0.1	0.1
Intermediate Results					
1. Heat duration time in the previous month, min	56.80	58.41	69.16	74.54	79.92
2. Heat duration time in the current month, min	58.41	58.41	58.41	58.41	58.41
3. Efficiency of the duration of the heat, I_1	0.97	1.00	1.18	1.28	1.37
4. Share of time duration arc under current in the previous month	0.52	0.53	0.58	0.60	0.63
5. Share of time duration arc under current in the current month	0.53	0.53	0.53	0.53	0.53
6. Efficiency of share of time duration arc under current, I_2	0.99	1.00	1.10	1.14	1.19

* – planned constant value of the proportion of heats with the third scoop in the current month, no more 10 %.

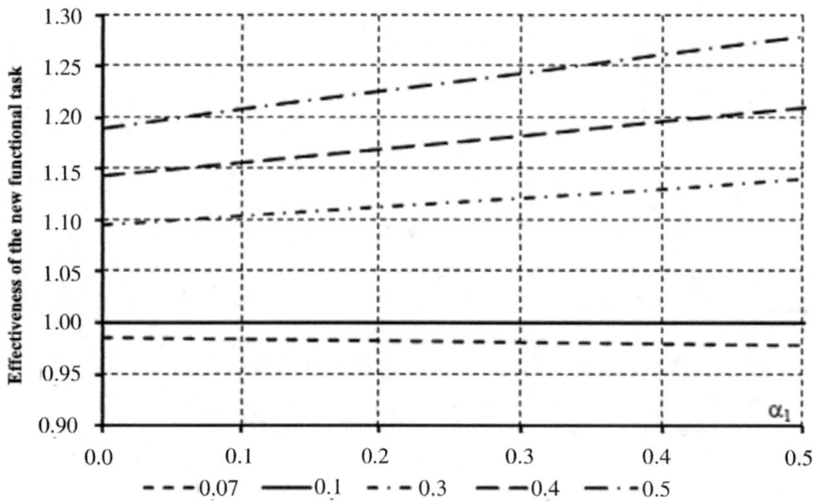

Figure 4. The results of determining the total effectiveness of the new functional task.

4) The possibility of reducing the number of scoops in charging process with the new functional task module was presented.

References

[1] Lapshin I V 2004 *Arc Furnace Automation* (Moscow: Moscow State University Press) p 165
[2] Trukhov A P and Malyarov A I 2004 *Foundry alloys and smelting* (Moscow: Moscow State University Press) p 336
[3] Parsunkin B N, Usachev M V, Logunova O S and Koroleva V V 2019 *Journal of the Minerals Metals & Materials Society (JOM)* **71** pp 342–48

IOP Conf. Series: Materials Science and Engineering **718** (2020) 012011 doi:10.1088/1757-899X/718/1/012011

[4] Mordasov D M and Mordasov M M 2004 *Technical measurements of the density of bulk materials* (Tambov: TGU) p 80
[5] Li Y, Pan R and Zhu S 2019 *Archive for Rational Mechanics and Analysis* **234** pp 1281–1334
[6] Logunova O S and Oshurkov V A 2019 *Cherepovets State University Bulletin* **2** pp 32–43
[7] Oshurkov V A, Egorova L G, Lednov A V and Antipanov I D 2019 *Electrotechnical Systems and Complexes* **1** pp 59–66
[8] Logunova O S, Sibileva N S and Pavlov V V 2016 *Steel in Translation* **10** pp 733–738
[9] Logunova O S, Oshurkov V A, Sibileva N S, Gavritskov S A and Garbar E A 2019 *Steel in Translation* **4** pp 22–25
[10] Logunova O S, Romanov P U and Il'ina E A 2019 *Processing of experimental data on a computer* (Moscow: Dom «Infra-M») p 326

IOP Conf. Series: Materials Science and Engineering **718** (2020) 012012 doi:10.1088/1757-899X/718/1/012012

On the Applicability of the Evaluation Techniques of the Properties of Iron-Ore Sinter

A V Malygin[1] and T V Detkova[2]

[1] LLC Ferrox, 36 Sovetskikh Zhenshchin street, Ekaterinburg, 620146, Russia
[1] PAO Severstal, 30 Mira street, Cherepovets, 162608, Russia

E-mail: a.malygin@ferrox.org

Abstract. Referring to the hierarchical structure of sinter cake, the applicability of various methods for controlling sinter properties has been studied. The technique has been proposed for evaluation of quality yield and strength for sinter using the parameters characterizing structural fines yield and strength of macroblocks under destruction.

With the development of blast furnace production, the requirements to the quality of sinter have been increasing continuously, in particular the requirements to its physical-chemical properties (size, cold and hot strength) [1-3]. Sinter cake formed as the result of melting and shrinkage of charge materials in difficult and unbalanced conditions [4-6], is structurally a largely inhomogeneous polymineral polycrystalline fractured porous body which is in an internally stressed state. This definition formulated by S. V. Bazilevich [4], to the fullest extent possible reflects the physical essence of sinter cake and predetermines the complicated multifactorial dependence of the properties of sinter on its composition and structure.

Despite long-term production experience and many scientific studies, the production of high-quality sinter meeting all the requirements is still a difficult but at the same time important and relevant objective of science and technology. In these circumstances the system of the evaluation of the properties of sinter is of great importance and, in particular, the search for new quantitative methods of the evaluation of its quality indicators is important.

Taking into account the multiple stage process of destruction and the volume inhomogeneity of the distribution of substance in sinter cake, its structure can be presented as a particular scale hierarchy [7]: the systems of interconnected structural elements of diversified scale level: sinter cake, its lump fragments, blocks and mineral aggregates, separate grains of minerals, etc. The highest structural level of the hierarchy (sinter cake) has the most complicated structure, because it is presented with all diversification of the lower ones and the connections among the elements of lower structural levels which make it up. Any structural element of sinter cake is an element of the scale hierarchy, its structure is simpler and it does not have all of the same properties, but at the same time it is a component part of the sinter cake. In accordance with the sequence of destruction of sinter cake and the formation of consumer properties of sinter, 5 structural layers of the hierarchy can be identified depicted on Figure 1.

Every structural element of the hierarchy is responsible for the provision of a particular group of properties, corresponding to its scale. The structure of sinter cake must provide the optimal yield of sinter. The structure of lump products of crushing with the maximum size 100-200 mm should provide

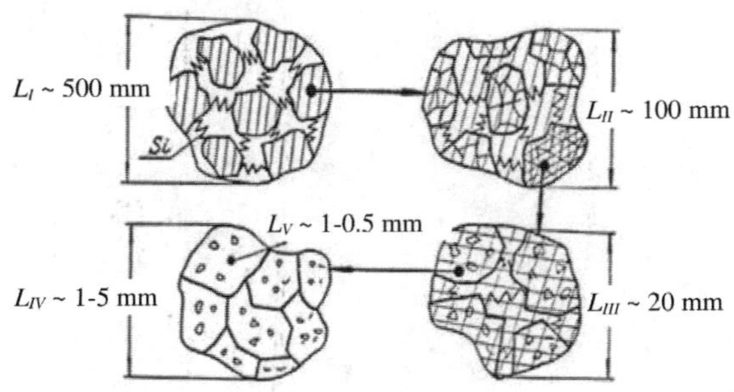

Figure 1. The hierarchy of the sinter cake structure: L – the size of a structural element; S_i – the surface of destruction of connections among elements.

the acquisition of lumps size from 5 to 40 mm (blocks) with the minimum formation of fines 0-5 mm. The structure of lumps of size 5-40 mm should provide minimum destruction during both mechanical impact and restorative heat treatment, and also the lumps should have high physical-chemical properties. The structure of sinter cake is a basic parameter which reflects the results of sintering and destruction. The analysis of the influence of the parameters of sintering process on the yield and mechanical strength of sinter by the microstructure of the lumps of yield stabilized sinter bypassing the complete macrostructure of the sinter cake is inefficient.

In accordance with the aforementioned, the existing methods of destructive control of yield and the strength of sinter are essentially correct, but they have significant drawbacks. In industrial conditions the yield is considered to be everything that is dispatched to the blast furnace shop or (at some enterprises) the yield of skip sinter. In these circumstances the yield sinter can include a significant amount of small (0-5 mm) and large (more than 40 mm) fractions depending on the efficiency of crushing and screening. The yield in research practice is considered to be the residual content of classes from +5 mm to +12mm in the products of destruction of sinter cake after 1-3 drops onto a steel plate from 2 m height.

The indicator of the strength of sinter, as a rule, is considered to be the residual quantity of the class +5 mm (+6.3 mm) or the yield of the class 0-5 mm after the destruction (test) of the lumps of yield sinter of a particular size in a rotating drum. As the indicator of sinter strength during restoration we can take also either residual content of initial size classes in a tested sample, or the output of over-extreme (fine) classes.

The specified methods of evaluation of yield and strength of sinter are to a great extent formal, less informative and do not allow us to make conclusions about the reasons of the destruction of sinter during tests. Since the grain-size composition of the products of destruction depends both on the structure received after sintering and on the parameters of the destruction process, with the help of the existing indicators we cannot estimate the reasons of the decrease or increase of yield and strength of sinter.

As theoretical and experimental studies showed [8], in the process of multiple stage destruction of sinter cake by impact, which is applied in research practice and exists in real industrial conditions, the mass fraction of fines 0-5 mm in the products of destruction is connected to the work of the destruction by the dependence:

$$g_m = g_o + C \cdot U^{0.5}, \quad (1)$$

where g_m is the mass fraction of fines in the products of destruction of sinter cake (%); g_o, C – coefficients; U is the specific work spent for the destruction of sinter cake (J/kg).

The coefficient g_o is not connected to the external mechanical energy spent for the destruction and depends on the macrostructure of sinter cake (total porosity, size, shape, specific surface area of interblock pores). The value g_o determines the percentage of structural fines, present in sinter cake as links between blocks, and also it reflects all kinds of incomplete sintering (insufficient sintering). The strength of “open” lumps of sinter during disintegration of sinter cake is characterized by coefficient “C”.

The equation (1) shows that the rate of incrementation of fines in the process of destruction is also connected to the work of destruction:

$$dg/dU = 0.5C \cdot U^{-0.5}. \quad (2)$$

If for the mechanical testing a sample of sinter is taken, which was preliminarily destroyed by the application of specific energy $U_2 = U_1 + \Delta U$, where ΔU is the energy spent on the destruction in a test, then the incrementation of fines in the process of testing (Δg) can be determined by integrating the equation (2) within the limits of the change of the work of destruction U_1 to U_2:

$$\Delta g = C(U_2^{0.5} - U_1^{0.5}). \quad (3)$$

Equation (3) shows that the incrementation of fines in the process of testing depends both on the energy spent on the destruction in the process of testing, and on the energy spent on preliminary destruction. That is why the results of strength determination by existing accepted methods depend on the place over the route of the transfer of sinter from a machine to a blast-furnace shop where the sample for testing was taken from, on whether the destructive impact is chosen, on a test and also on the value "C". In connection to that the determination of yield and strength of sinter by the existing formal indicators does not fully reflect the true results of sintering.

The scheme of technological analysis which allows to experimentally determine the coefficients g_o and C and, together with that, the group of factors responsible for the acquisition of particular physical properties of sinter cake, is quite simple. The laboratory cake or a fragment of industrial cake of the volume 0.03-0.05 m^3 is consecutively destroyed by multiple drops from a particular height (usually 2 m) onto a steel plate or by treatment in a rotating drum of 1 m diameter. In these circumstances the change of the output of small fractions (0-5 mm) is recorded depending on specific energy spent during destruction (minimum after three, five, seven and ten drops of cake from the height of 2 m onto a steel plate). By mathematical processing of the achieved experimental data in accordance with equation (1) the value of coefficients g_o and C is determined. The determined values of the coefficients are compared with the control ones and in case of deviation a decision is made regarding the application of a certain correcting measure.

The determination of coefficients g_o and C in industrial conditions is done by the control and analysis of the output of sintering products (return, yield sinter, the content of fines in sinter). Based on the results of control, the dependence of the output of fines 0-5 mm on the work of sinter destruction is determined, and the coefficients g_o and C are calculated. The achievement of the maximum content of fines in sinter is ensured by the correct ratio of g_o and C, it can be corrected in any particular case by the change of the preparation of charge and the mode of sintering.

The methods of non-destructive evaluation of strength properties of sinter are known. Thus, the authors of [9] considering sinter as an object of a complicated mineral composition with physical-mechanical properties characteristic for every phase, proposed the following equation of cold strength (SI) of sinter:

$$SI = K_{SI} Q \sigma_0 (1-\varepsilon)^2, \quad (4)$$

where Q is the amount of melt which connects the phases of sinter (the degree of melting); σ_0 is the strength of the matrix; ε is the porosity; K_{SI} is the coefficient of proportionality. Values Q, σ_0 and ε are determined by calculation.

Low accuracy of the determination of the strength indicator of sinter per the specified model (the coefficient of correlation R=0.675) is connected, firstly, with the use of the properties of a low level element of the hierarchy (σ_0) for the determination of strength of a high level element. The strength of the matrix (σ_0) is connected not only to its mineral composition, but also with its structure, in particular, with the parameters of the distribution of material in sinter, grain size, porosity and the character of porosity of the mineral skeleton.

In [10] the information about the close connection of the strength indicator of sinter with its chemical composition is given through a so-called module of microstructure:

$$M_m = \frac{Fe_{gen}}{FeO} \frac{CaO}{SiO_2} . \tag{5}$$

Based on the data of the author, the strength indicator of sinter of sinter plant in CIS per GOST 15137-77 is basically linearly ($R^2 = 0.879$) increasing with the increase of the module of microstructure, which contradicts the existing industrial and experimental results. If we consider that the strength of sinter increases when Mm is increasing, then the higher the content of FeO in the sinter, the lower its strength, which is not consistent with reality. The dependence of the strength indicator of sinter on its basicity is non-linear. Besides, the cold strength of sinter depends on the macrostructure of sinter cake, its porosity and the character of porosity (first and second levels of the hierarchy). The module of microstructure reflects the properties of the lower elements of the structural hierarchy of sinter cake and cannot account for the cold strength of sinter.

The hot strength of sinter is closely connected to the parameters of the chemical composition of sinter [11], presented on Figure 2, that the lower elements of the hierarchy of sinter cake account for.

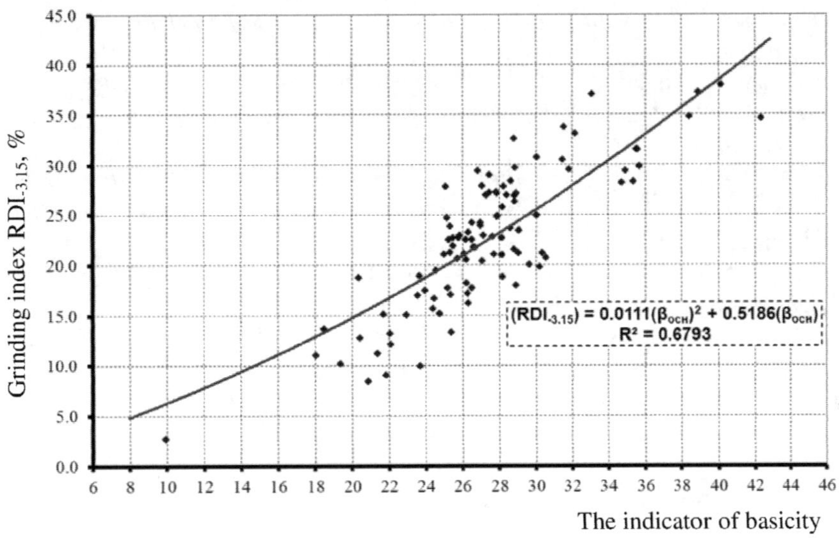

Figure 2. The influence of the indicator of basicity of sinter on the grinding after restoration $RDI_{-3.15}$ per ISO 4696-1.

Thus, the phase transition from hematite to magnetite influences the strength after restoration, which is accompanied by the change of parameters of the crystalline grid and the creation of internal stress [5]. The strength after restoration depends on the fraction both of primary and secondary hematite, the content of which increases when increasing the fraction of FeO and SiO_2 in sinter and increasing Al_2O_3 content. Al_2O_3 can also cause disruptions in the magnetite grid, which is created after the restoration of hematite and it can increase the internal stress. The basicity can influence the strength after restoration, also the fraction of MgO in sinter and other factors [12]. The influence of the chemical composition on the strength of sinter after restoration is quite significant, which allows us to use the dependencies determined in [11] for the analysis and the prognosis of its hot strength.

Conclusions

Thus, for the evaluation of yield and the indicator of the mechanical strength of sinter it is reasonable to use the methods based on the destruction and control of the grain-size composition of the products of destruction of sinter cake. The most informative method is the one which allows evaluation of the fraction of structural fines in sinter cake and the indicator which reflects the strength of the blocks. The connection of the properties of sinter with its chemical composition becomes significant only when evaluating its strength after restoration.

References

[1] *Iron metallurgy: textbook for higher education* 2004 Ed. Yu S Yusfin (Moscow: IKC "Akademkniga") p 774
[2] Kortman H, Lungen G-V, Jacobs V and Gudenau H V 1988 *Chernye metally* **21** 29–36
[3] Malygin A V, Maltsev V A and Viduyetskiy M G 2016 *Ore preparation processes in melting production* (Ekaterinburg: UrFU) p 415
[4] Bazilevich S V and Vegman E F 1967 *Agglomeration* (Moscow: Metallurgy) p 368
[5] Vegman E F 1974 *The theory and technology of agglomeration* (Moscow: Metallurgy) p 288
[6] Korotich V I, Frolov Yu A and Kaplun L I 2009 *The theoretical basics of the technology of lumping of metallurgical raw materials. Agglomeration* (Ekaterinburg: GOU VPO UGTU-UPI) p 416
[7] Malygin A V 1999 *The scientific basics and the practice of improvement of the process of iron-ore sinter production with high customer properties: doctoral dissertation* (Ekaterinburg: UGTU-UPI)
[8] Malygin A V and Shumakov N S 1997 *Izvestiya vuzov. Iron and steel metallurgy* **9** 9–12
[9] Itidate M et al. 1988 *Mathematical modeling of the influence of major factors on the cold strength of sinter* (Sverdlovsk: VNIIMT)
[10] Nechkin G A 2015 *The study of the influence of mineral structure formation on the metallurgical properties of iron-ore sinter: dissertation in support of candidature for a technical degree* (Ekaterinburg) p 142
[11] Eliseev A A, Detkova T V and Kuchin V Yu 2016 *Ferrous Metallurgy: Bulletin of Scientific, Technical and Economical Information* **2** 30–33
[12] Katsuhiko S, Satoru S, Yasumasa S and Katsuya O 1982 *J. Iron and Steel Inst.* **68** (15) 2215–2222

Analysis of the Microstructure Evolution in the Implementation of a New Macro Shear Rolling Technology

A B Naizabekov[1], S N Lezhnev[1] and E A Panin[2]

[1] Rudny Industrial Institute, 50 Let Oktyabrya Street 38, Rudny, 111500, Kazakhstan
[2] Karaganda State Industrial University, Republic Avenue 30, Temirtau, 101400, Kazakhstan)

E-mail: sergey_legnev@mail.ru

Abstract: In the work on the basis of computer modeling, a comparative analysis of the rolling technology of a plate billet, which includes rolling in relief rolls with subsequent rolling in rolls with a smooth barrel to the desired size, compared with the existing rolling technology was performed. The results of comparison of deformation schemes using microstructure evolution indicate the advantage of the proposed technology compared with the current rolling technology.

The quality of sheet products depends on the mechanical properties of the material, and, consequently, the uniformity of distribution and dispersion of the granulated structure. It was noted in the works [1, 2] that the rolled structure formed after rolling through roughing stands – has a great impact on the further processes of structure formation and the formation of the final structure of a plate. At this stage, the control of the processes of structure formation pursues two main objectives: 1) obtain a fine austenite grain and 2) prevent the formation of a mixed structure of partial recrystallization. Moreover, it is known from works [3, 4] that, during reduction of a slab, there is a significant unevenness of deformation in the volume of the body being deformed - surface layers are more deformed, while the middle part of the strip is barely or not deformed at all, and this can lead to the formation of a heterogeneous structure. It is also known that the formation of the structure of heterogeneous austenite after rolling through roughing stands causes the formation of a heterogeneous ferritic structure and cannot be corrected during the rolling through finishing stands [4].

It is also known that it is impossible to completely eliminate the unevenness of deformation due to the monotonous nature of the metal flow, as a result of which, when rolling high strips, the molten structure of the slab's axial zone is not sufficiently worked out, nonmetallic inclusions and liquids have nodule alignment in the metal along the rolling axis, which reduces the required level of physical and mechanical properties of products.

A promising solution to this problem is the development of new billet deformation schemes, which allow creating additional metal flows when rolling, i.e. to intensify the developed shear deformations not only in the longitudinal but also in the transverse direction.

Intense macro shears can be provided in a variety of technological and constructive ways. The most known and promising among them are the following: asymmetric rolling [5, 6]; the use of crossed rolls [7]; the use of rolls and billets with a corrugated or wavy surface [8–10]; the use of rolls with surface asperities [11, 12]. All of the above mentioned methods have one thing in common - intense macro shears are achieved as a result of local deformation action on the rolled metal.

One of these technologies is the technology of rolling a plate, which includes the following: the use of rolls with a grooved surface [13], followed by rolling through the smooth rolls to the required size. The surface of the grooved rolls is made in the form of annular grooves forming a trapezoidal shapes along the entire length of the roll which are located at an angle of 90° to the rolling axis. When rolling through grooved rolls, trapezoidal segments are introduced into the body of the billet and due to the peculiarities of their shape, a part of the metal is displaced in-between these segments. As a result of this, shear deformation is intensified along the billet cross section with the formation of trapezoidal segments on the billet surface. During subsequent rolling of the billet in smooth rolls, conditions are created to ensure an alternating metal flow when leveling the transfer bar surface while maintaining the original geometry of the billet.

The objective of this work is to analyze the effectiveness of the proposed technology for rolling a plate, in comparison with the existing rolling technology (smooth rolls), based on the analysis of the evolution of the microstructure using computer modeling.

To build geometric models of the billet and tool, the KOMPAS software was used, in which 3D-models were created and exported to the compatible STL format. As a result, after importing the geometry files into the Simufact Forming software, a computer model was obtained (Figure 1), consisting of 3 pairs of rolling stands in series. The rolls of the first stand are grooved, in which the billet undergoes shaping and shear deformation. The second and third stands are equipped with smooth rolls, which serve to flatten the deformed billet in order to return its original geometric parameters.

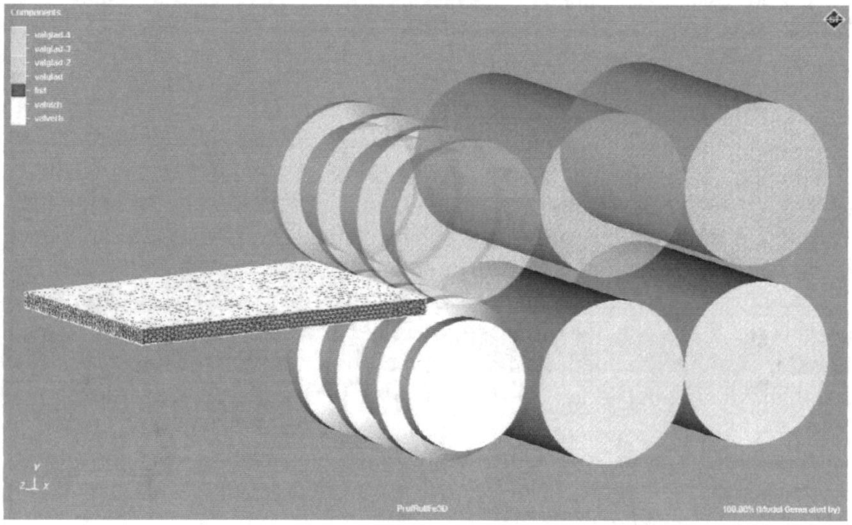

Figure 1. Plate rolling process model using new power saving technology.

Based on previous works [14], when modeling a new technological rolling scheme, including rolling through grooved rolls, it was decided to use grooved rolls with an unequal up to down ratio, ensuring the same roll gap at different points.

The use of such rolls makes it possible to implement a simple shear scheme when rolling, which most favorably affects the preservation of the initial dimensions of the billet than the use of rolls with an equal ratio of up to the down, where, in addition to shear, reduction is also carried out on inclined sections of the rolls.

The billet is a rectangular sheet with the following dimensions: h x b x l = 10 x 140 x 200 mm. The material used for billet is brass of L63 brand.

In computer model of the process, the following technological parameters were used:

IOP Conf. Series: Materials Science and Engineering **718** (2020) 012013 doi:10.1088/1757-899X/718/1/012013

- The rolling was performed at a room temperature of 20 °C;
- The billet temperature before rolling is 600°C;
- Heat conductivity coefficient is $7000\, W/(m^2 \cdot °C)$;
- Siebel friction model, i.e. contact stress exceeds yield strength;
- Friction coefficient is 0,7;
- Rolling speed is 1,25 rad/s.

Rolling process can be divided into three main stages (Figure 2).

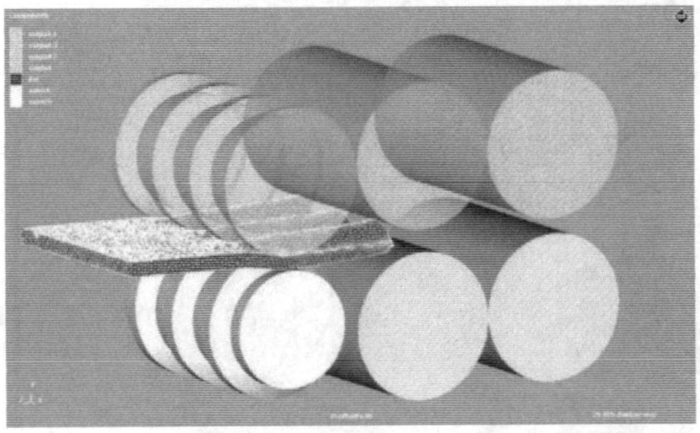

a

b

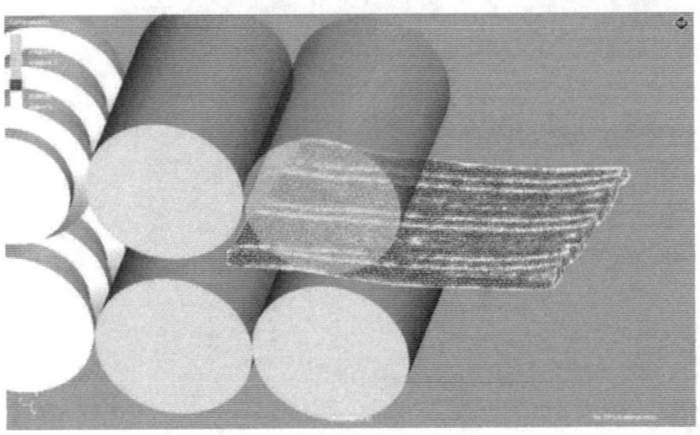

c

Figure 2. Rolling process stages:
(a) – deformation in the grooved rolls;
(b) – deformation in the first pair of smooth rolls;
(c) – deformation in the second pair of smooth rolls.

At the first stage, the billet preheated to the rolling temperature is fed into the grooved rolls and during the first pass a single reduction by the rolls' collars is carried out until the groove cavity is completely filled with metal (Figure 2 (a)). After rolling in the first stand, alternating ups and downs in the form of trapezoidal segments are formed on the surface of the billet. This stage is mainly characterized by shear deformation, however, there is also a height deformation at the junction of the ups, which contributes to the capture of the billet. In order to pre-flatten the surface of the profiled billet after rolling in the grooved rolls, rolling through the smooth rolls is carried out (Figure 2 (b)). As a result, conditions are created to ensure an alternating metal flow when flattening the surface of the transfer bar during rolling through smooth rolls, maintaining the original shape of the billet. In the third pass, the billet was also fed through the stand with smooth rolls (Figure 2 (c)).

Figure 3 presents the results of modeling the microstructure, the initial grain size is 60 μm (Figure 3 (a)). When rolling the billet through the grooved rolls (Figure 3 (b)), a significant refinement of the microstructure due to shear deformation is observed, the grain size reaches 35–45 μm.

From the cross section of the billet at the second stage (Figure 3 (c), it can be seen that the sections with the finest structure are the ups of the billet, the grain size is about 25–35 μm.

IOP Conf. Series: Materials Science and Engineering **718** (2020) 012013 doi:10.1088/1757-899X/718/1/012013

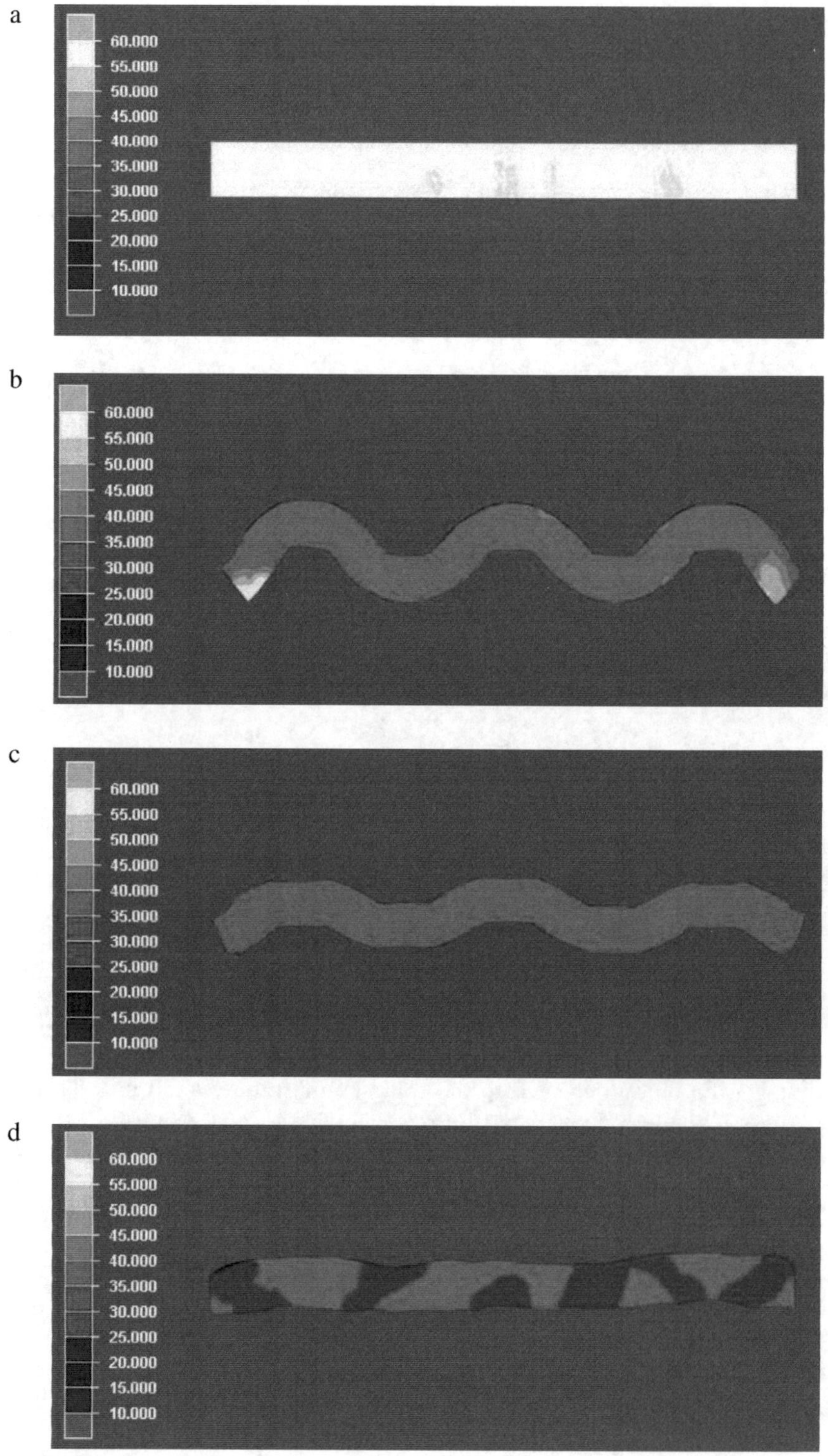

Figure 3. Evolution of the microstructure (grain size) when rolling: (a) – initial billet; (b) – the billet after grooved rolls; (c) – the billet after second pass (smooth rolls); (d) – the billet after third pass (smooth rolls).

IOP Conf. Series: Materials Science and Engineering **718** (2020) 012013 doi:10.1088/1757-899X/718/1/012013

This is due to the fact that the ups are the points of deformation concentration. In other parts of the billet the grain size varies from 40 to 30 μm (mkm).

At third pass stage (Figure 3(d)) the grain size reaches 30–25 μm. At the base of the ups the grain size is 25-20 μm.

For comparative analysis, two models were created and modeling was carried out using the Simufact Forming software. For the current technology the production technology of a thick sheet (made of copper alloys) at the Balkhash non-ferrous metal processing plant is taken. The model of the existing technology for rolling a plate includes 6 passes in a DUO-100 stand with smooth rolls. In our case, for the convenience of modeling, 6 passes of the billet through the DUO-100 stand were replaced by rolling in 6 consecutive stands with smooth rolls. The model of the proposed technology includes a stand with grooved rolls in the form of annular grooves along the entire length of the roll and 5 sequentially installed stands with smooth rolls.

The peripheral rolling speed in the stands was calculated from the condition of constant second volume (amount of metal going through the stand per second) in order to stretch the strip and ensure horizontal passage of the billet through the stands, as well as to exclude loop formation when rolling. The peripheral rolling speed in the stands is presented in Table 1.

Table 1. Peripheral rolling speed (per stand).

Technology	Stand 1	Stand 2	Stand 3	Stand 4	Stand 5	Stand 6
Existing, rpm	30	35.42	40.48	43.97	47.2	51
Proposed, rpm	30	40	45	60	81.8	90

It is known that the temperature of the billet is an important technological parameter. The calculation results of the change in the temperature of the billet in each pass in each of the compared models are presented in Figure 4.

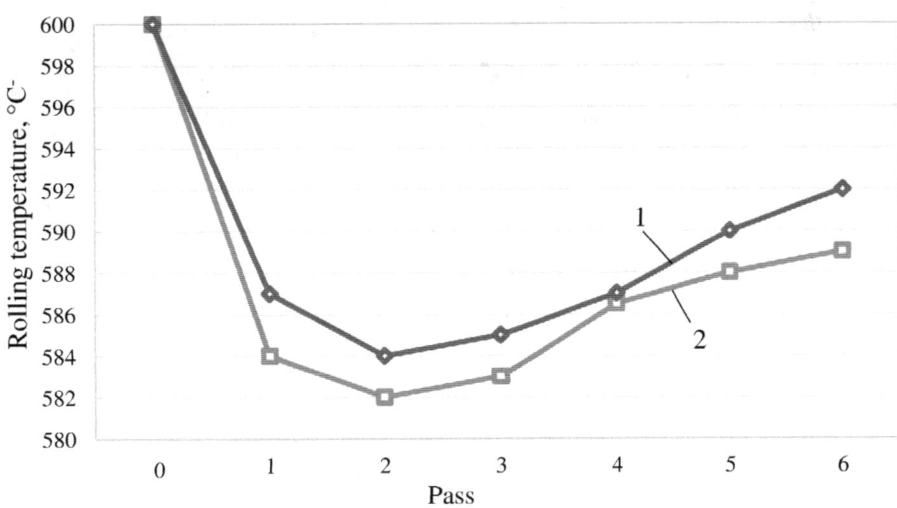

Figure 4. Billet temperature change graph (per pass):
1 – existing technology; 2 – proposed technology.

From this graph it follows that the billet rolled by the proposed technology cools down more than by the existing technology. This is due to the fact that when rolling according to new technology in the first three stands, there is no significant reduction of the billet, but only a change in its form and flattening.

In the last three stands, an increase in temperature is observed in both models. This phenomenon is explained by an increase in the rolling speed, as well as an increase in the degree of reduction.

In the framework of this model, a comparative analysis of the billet microstructure after rolling through the first stand was performed (Figure 5 (a)), which revealed that when rolling through a grooved stand, the microstructure has more even distribution and a smaller grain fraction than when rolling through the stand with smooth rolls. This is due to the large value of compressive strain in the proposed technology. The grain size reaches a value of 12–14 μm, while in the existing technology 14–16 μm.

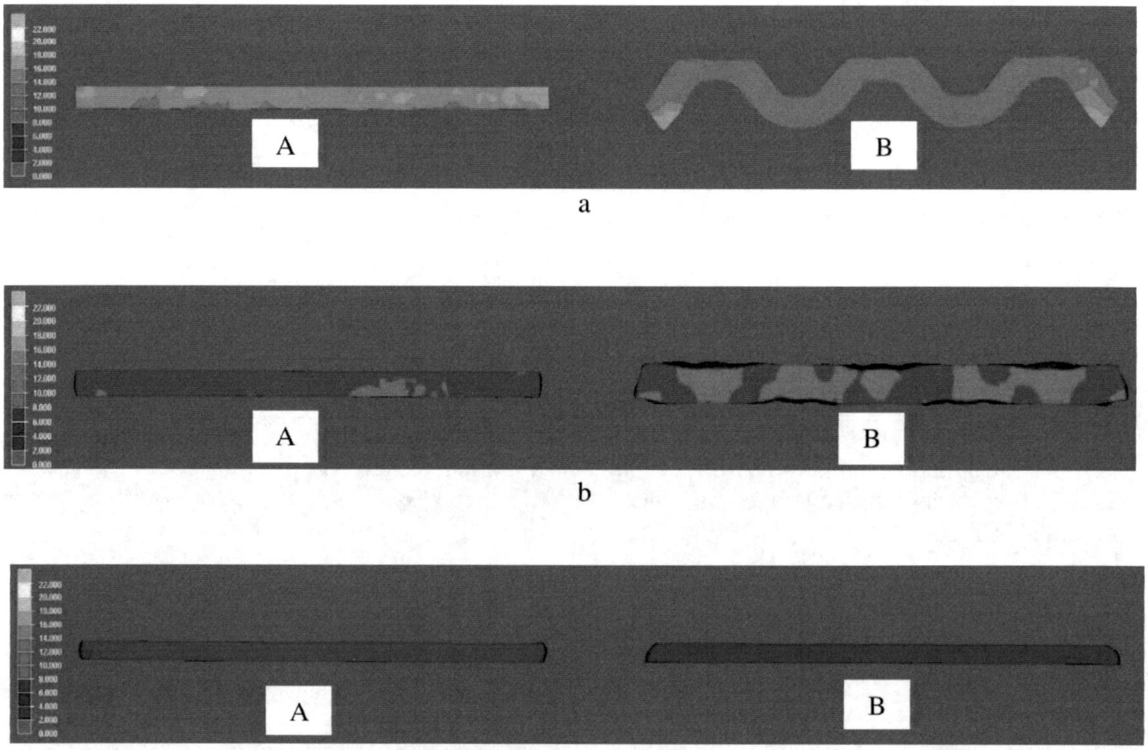

Figure 5. The results of modeling the evolution of the microstructure: (a) – after rolling through the 1st stand, (b) – after rolling through the 3rd stand, (c) – after rolling through the 6th stand; A – existing technology; B – proposed technology.

After the last smoothing stand (the third one), a more uniform distribution of the structure is observed in the rolled billet by the existing technology (Figure 5 (b)). The grain size in both models reaches 8 μm. In billet "B", the smallest grain size prevails near the grooves.

Figure 5 (c) shows the microstructure after the 6th stand. The billet "B" has a uniform distribution of structure with a grain size of 2–4 μm. The grain fraction in the billet "A" ranges from 5 to 8 μm, the microstructure has an inhomogeneous structure, which leads to anisotropy of the properties. It can be assumed that the use of the proposed rolling technology will help to avoid additional (excessive) metal loss in the form of side trimmings in order to equalize the mechanical properties of the sheet.

Conclusion

A comparative analysis of two deformation schemes: the proposed one, which includes rolling through grooved rolls followed by rolling of a billet through smooth rolls, and the current one (rolling only through the smooth rolls). Analysis results indicate the advantage of the proposed deformation scheme compared to the existing one. Thus, the use of the proposed scheme of deformation in the first stand (with grooved rolls) contributes to a more intensive grinding of the initial grain size when rolling in all six stands.

References

[1] Tanaka T, Tabata N, Hatomura T and Shiga C. 1977 *Micro alloying 75. Proc. Intern. Symp.* (New York: Union Carbide Corp.) pp 88–99
[2] Irvine K J, Pickering F B and Gladman J J 1970 JISI **208** 717–726
[3] Pogorzhelsky V I 1986 *Controlled rolling of continuously cast metal* (Moscow: Metallurgy) p 151
[4] Efron L I 2012 Metal science in the «big» metallurgy. Pipe steel (Moscow: Metallurgizdat) p 696
[5] Piesin A M, Salganik V M, Dyja H, Chikishev D N, Pustovoitov D O and Kawałek A 2012 *Hutnik Wiadomości Hutnicze* **5** 358–362
[6] Isaevich L A, Korol' V A, Karpitsky V S, Sidorenko M I. and Ivanitsky D M 2010 *Casting and metallurgy* **4** 117–121
[7] Brinza V V and Odinaev B N 2000 *News of Higher Educational Institutions. Ferrous metallurgy* **7** 42–44
[8] Meandrov L V, Vorontsov V K, Pogorzhelsky V I and etc.1982 *Steel* **3** 43–45
[9] Brinza V V, Demchuk N N and Yarmak G M 1996 *News of Higher Educational Institutions. Ferrous metallurgy* **11** 29–33
[10] Dolzhenkov F E, Sunyaev A V, Ostapenko A L and etc. 1984 *Steel* **3** 33–37
[11] Bogatov A A, Nukhov D Sh and Pyankov K P 2015 *Metallurgist* **2** 14–16
[12] Mashekova A S, Kavalek A, Turdaliev A T and etc. 2016 *Bulletin of the National Academy of Sciences of the Republic of Kazakhstan* **2** 17–27
[13] Naizabekov A B and Lezhnev S N 2007 Rolls for rolling thick sheets *Patent of the Republic of Kazakhstan* No. 14791
[14] Naizabekov A B, Lezhnev S N and Ashkeev Zh A 2001 *News of Higher Educational Institutions. Ferrous metallurgy* **2** 25–26

Research on the Effect of Reduction Distribution in Universal Stands on the Process of Metal Transition from Sides to the Center Line of Transfer Bar

A V Polyakov[1] and I P Mazur[1]

[1] Lipetsk State Technical University, 30 Moskovskya Street, Lipetsk, 398043, Russia

E-mail: pool_akov@mail.ru

Abstract. The edge defects formed on the continuously cast slab before entering the hot rolling mill in the roughing group of the broadband mill are moved from the side faces and edges of the slab to the upper and lower surface of the slab. This process is facilitated by alternating compression in vertical and horizontal rolls. This article describes the transition of a metal from the edges and side faces of the slab at the top and bottom surfaces of the slab sheet during rolling in universal stands of the roughing stands of the hot rolling mill. The amount of movement of the defects on the surface of the slab directly affects the amount of lateral trim is necessary for assured removal of marginal defects in subsequent processing. The presented mathematical model based on the finite element method includes compression of continuously cast slab in three separate stands consisting of vertical and horizontal rolls. The influence of a number of technological parameters and modes of adjustment of the vertical rolls on the value of the metal displacement to the center line of the transfer bar is studied.

1. Introduction

In the production of steel strip, a significant number of jams, during both cold and hot rolling, happens due to the presence of defects of the strip, such as sliver, rolled crack, etc. Jam in the rolling mill due to strip breakage results not only in rejects, but also in mill downtime and roll damage. The most critical from the point of view of jamming defects are located near the edges of the strip.

There are several reasons for the formation of cracks, one of which is steel casting on curvilinear continuous-casting machines [1]. Furthermore, when rolling through the roughing stands of continuous wide-strip hot mill and spreading the strip, metal areas are moving from edges and sides to the horizontal surfaces [2, 3].

The objective is to assess the amount of metal transition from sides to the upper and lower surfaces of the transfer bar when rolling through the roughing stands of a continuous wide-strip hot mill, and also to determine the technological parameters of rolling that have the greatest influence on the metal transition. This problem was solved by conducting experiments with a help of a mathematical model based on the finite element method using the ABAQUS software.

2. Model description

The model includes 3 vertical and 3 horizontal rolls arranged in pairs (Figure 1). In this case, the pair of vertical and horizontal rolls form a continuous group (the transfer bar is simultaneously in the vertical and horizontal rolls), and the pairs are arranged in such a way that the transfer bar goes into the next pair

Figure 1. The model of rolling ¼ of slab in three universal roughing stands.

only after it leaves the previous one. Thus, we obtain a model of reduction of a slab in three universal roughing stands.

For the following modeling conditions:

1) slab's steel brand Ст3;

2) slab's temperature 1200 °C;

3) slab's dimensions: $H = 250$ mm; $B = 500$ mm; $L = 500$ mm;

4) friction coefficient between slab and rolls is constant and equal to 0,5 [4].

The following assumptions are made [4]:

1) process under review – symmetrical in width and thickness of the slab, isothermal;

2) rolls are non-compressible;

3) deformable medium is viscoplastic;

4) slab's material – homogeneous, isotropic, resistant to deformation described by the equation in accordance with the work [5]:

$$\sigma=[b-(b-a)e^{-B\varepsilon}][1-A_1(1-e^{-\lambda_1 t})-A_2(1-e^{-\lambda_2 t})-A_3(1-e^{-\lambda_3 t})] \quad (1)$$

where ε and t – deformation degree and time from the start of the process; a, b, A_1, A_2, A_3, B, λ_1, λ_2, λ_3 – coefficients depending on steel grade and metal temperature.

The diameters of the rolls are shown in Table 1.

Table 1. The diameters of rolls used in the model for rolling a slab in three universal roughing stands.

Type of the roll	Roll diameter, mm		
	Stand No. 1	Stand No. 2	Stand No. 3
Vertical	1000	800	800
Horizontal	1350	1080	1080

In the research of the transition of metal from the sides to the upper and lower surfaces, 3 points lying on the lateral edge of the slab are considered:

1. Point A – 50 mm from the «head» of the slab.
2. Point B – in the middle of slab's length (250 mm from «head» and «tail» of the slab).
3. Point C – 50 mm from the «tail» of the slab.

In this model, the reduction of the slab in three stands is considered, because it is in them that the greatest absolute reduction occurs and, as a result, the greatest spreading of the transfer bar. The inclusion in the framework of the model of the remaining roughing stands (usually there are 5-6 roughing stands) would lead to a significant complication of the model and an increase in the resources used without a drastic change in the results.

3. Modeling results

With the reduction mode No. 1, shown in Table 2, all the points under consideration moved to the upper edge of the transfer bar. The position of these points is shown in Figure 2.

Table 2. The modes of reduction (as per stand) in the model of rolling a slab in three universal roughing stands.

Reduction mode	Parameter under consideration	Stand No.1		Stand No.2		Stand No.3	
		VV	HV	VV	HV	VV	HV
No.1	Width/thickness after the stand, mm	450	180	450	144	450	96
	Local absolute reduction, mm	50	70	18*	36	12*	48
	Local percent reduction, %	-	28	-	20	-	33
No.2	Width/thickness after the stand, mm	460	180	470	144	470	96
	Local absolute reduction, mm	40	70	6*	36	8*	48
	Local percent reduction, %	-	28	-	20	-	33
No.3	Width/thickness after the stand, mm	440	180	430	144	430	96
	Local absolute reduction, mm	60	70	30*	36	16*	48
	Local percent reduction, %	-	28	-	20	-	33
No.4	Width/thickness after the stand, mm	410	180	435	144	440	96
	Local absolute reduction, mm	90	70	0*	36	0*	48
	Local percent reduction, %	-	28	-	20	-	33
No.5	Width/thickness after the stand, mm	474	180	454	144	440	96
	Local absolute reduction, mm	26	70	34*	36	30*	48
	Local percent reduction, %	-	28	-	20	-	33

Note. VV – vertical rolls; HV – horizontal rolls; * – the indicator is given taking into account spreading of the strip after going through the horizontal rolls with an established rolling mode.

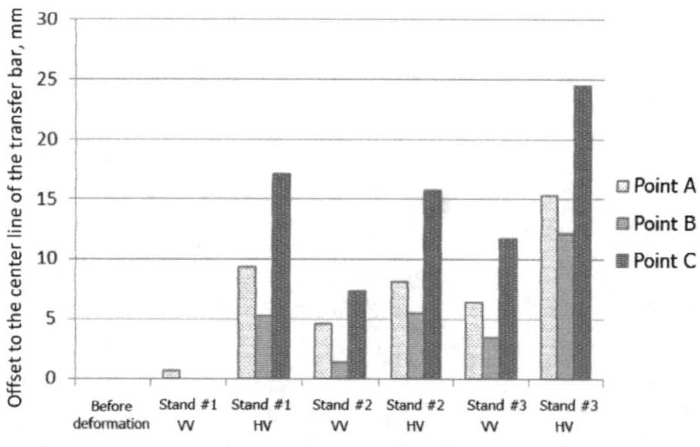

Figure 2. The offset of points to the center line of the transfer bar as it goes through the stands.

As expected, the smallest transition to the center line of the transfer bar turned out to be at a point lying on a section with an established rolling mode (B). After rolling in three stands (total percent reduction in thickness of 62%), point B moved by 12.1 mm (Figure 3). The greatest transition to the center line of the transfer bar was the point lying on the tail of the slab (C). After rolling in three stands, point C moved by 24.5 mm (Figure 3).

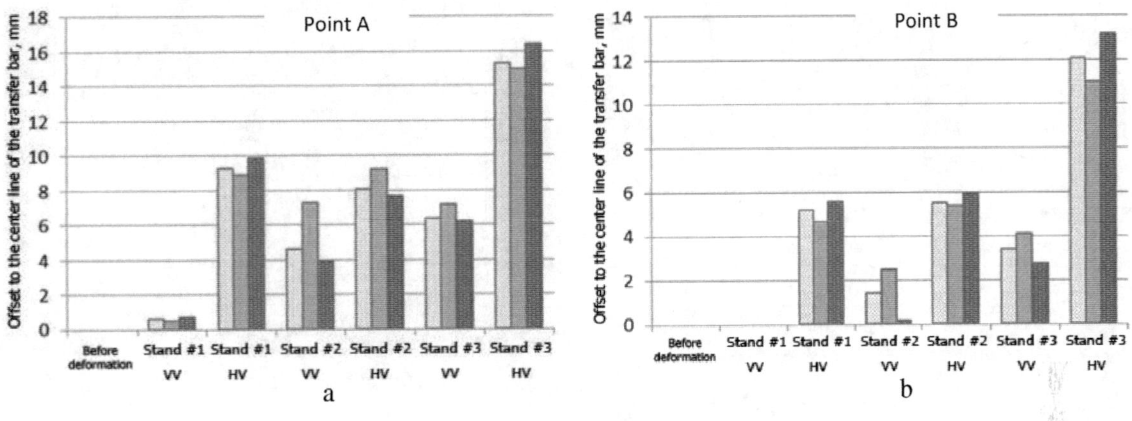

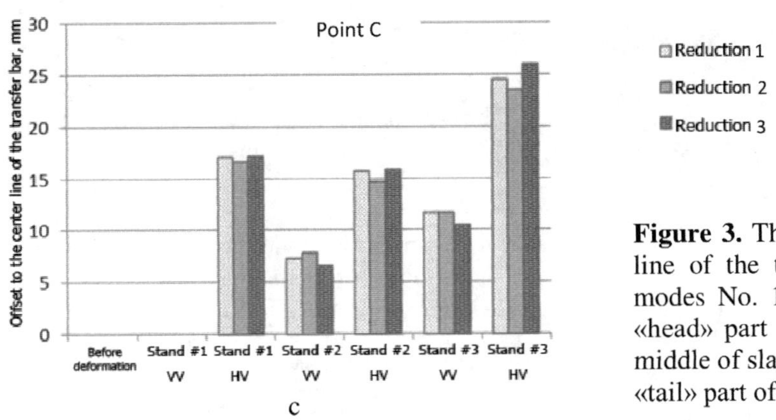

Figure 3. The offset of the points to the center line of the transfer bar when using reduction modes No. 1, 2 and 3: (a) – the point on the «head» part of slab (A); (b) – the point in the middle of slab's length (B); (c) – the point on the «tail» part of slab (C).

3.1. The effect of the edge reduction value

To assess the effect of reduction in vertical rolls, modeling was performed with a smaller (mode No. 2) and a larger (mode No. 3) absolute reduction in three stands relative to mode No.1, considered earlier.

A comparison of the transition value of the points under consideration to the center line of the transfer bar is shown in Figure 3.

All considered points moved mostly to the center line of the transfer bar with a greater width reduction (reduction mode No. 3).

3.2. The effect of reduction distribution (as per stand)

In this series of experiments, an additional limitation is introduced. The final slab width (after Stand No. 3) should be 450 mm.

Modeling was performed with heavy reduction in the vertical rolls of Stand No. 1 and the absence of reduction in the vertical rolls of Stands No. 2 and No.3 (mode No. 4), as well as with uniform reduction in the vertical rolls of stands No. 1 to No. 3 (mode No. 5).

The offset of the points to the center line of the transfer bar at various reduction modes is shown in Figure 4.

IOP Conf. Series: Materials Science and Engineering **718** (2020) 012014 doi:10.1088/1757-899X/718/1/012014

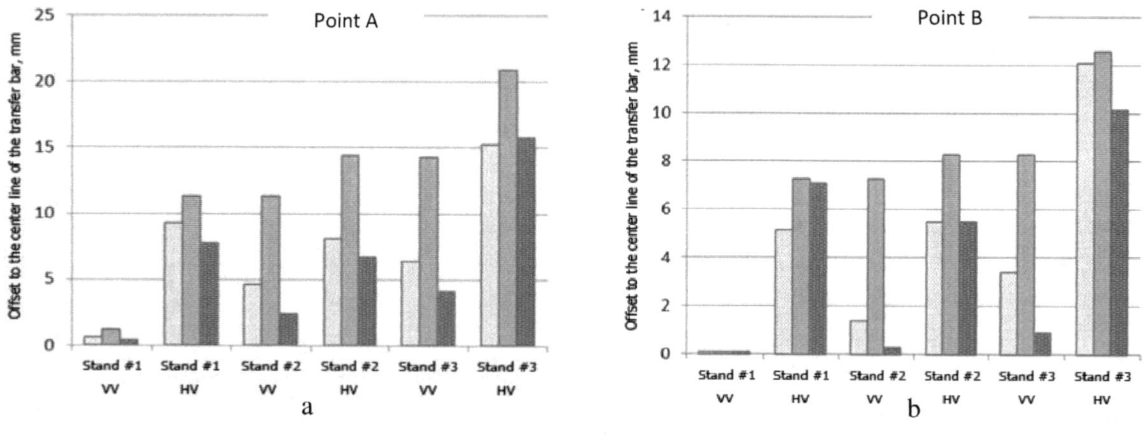

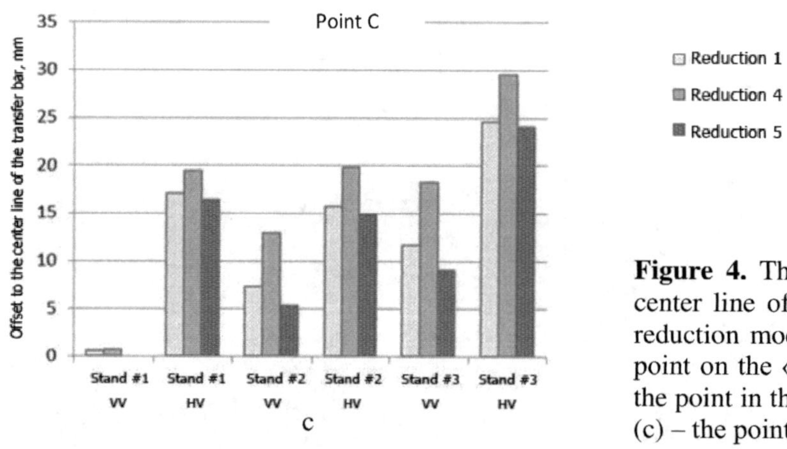

Figure 4. The offset of the points to the center line of the transfer bar when using reduction modes No. 1, 4 and 5: (a) – the point on the «head» part of slab (A); (b) – the point in the middle of slab's length (B); (c) – the point on the «tail» part of slab (C).

All metal located on the edge of the slab received the greatest reduction when using mode No. 4, corresponding to excessive width reduction in the vertical rolls of stand No.1 and the formation of the necessary strip width due to broadening during horizontal reduction. In this case, the vertical rolls of stands No. 2 and No. 3 serve as guiding rulers (reduction does not occur).

When assessing the minimum transition to the center line of the transfer bar, point B is taken (the point in the middle of slab's length). Its transition is minimal under mode No. 5, which corresponds to uniform width reduction in all stands. Points A and C in modes No. 1 and No. 5 have transitioned by a comparable amount.

Thus, the preferred mode of width reduction, from the point of view of minimizing the transition of metal from sides and edges of the slab to the center line of the slab, is a mode in which the absolute width reduction in all stands is distributed uniformly.

3.3. The effect of speed ratio of vertical and horizontal rolls

A different ratio of linear speeds of horizontal and vertical rolls forming a pair and its effect on the transition of the points to the center line of the transfer bar are considered. During the experiment, three cases were considered:

- Linear speed of vertical rolls is equal to linear speed of horizontal rolls of the corresponding stand;
- Linear speed of vertical rolls is 30% higher than the linear speed of horizontal rolls of the corresponding stand; thus, the vertical rolls support the horizontal rolls;

– Linear speed of vertical rolls is 30% lower than the linear speed of horizontal rolls of the corresponding stand; thus, the tension between vertical and horizontal rolls is created.

The modeling results are shown in Figure 5.

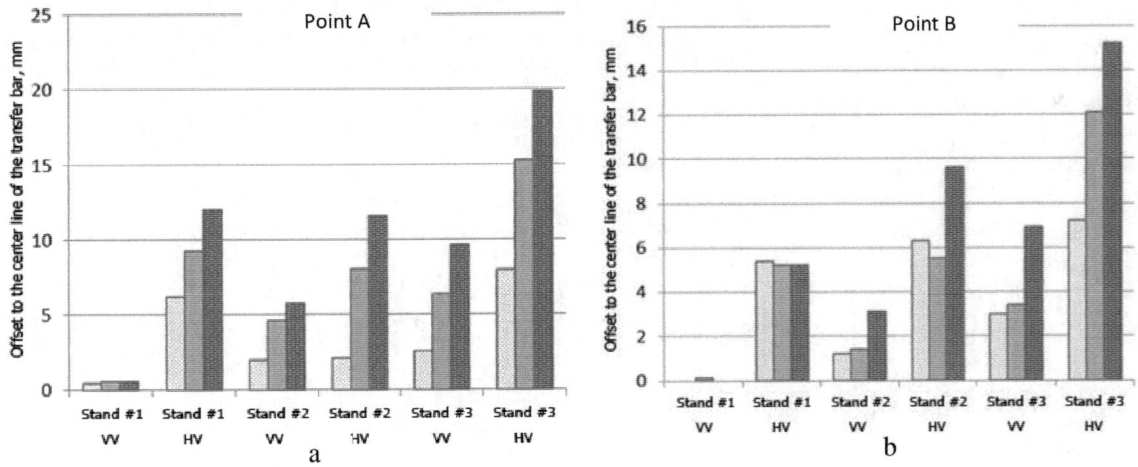

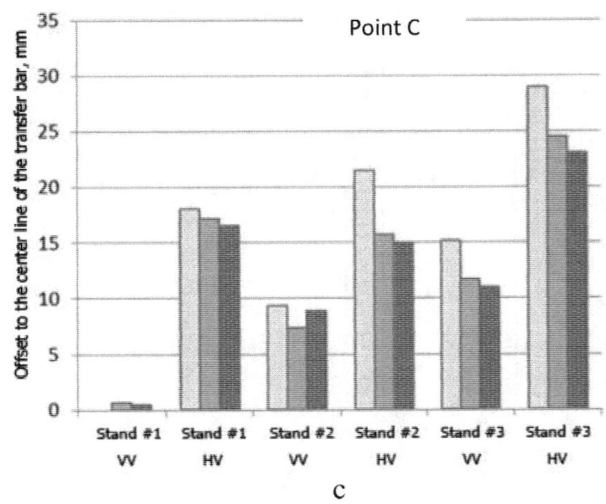

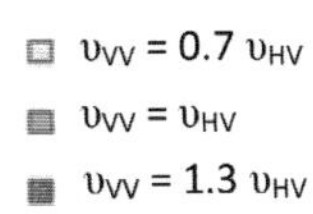

Figure 5. The offset of the points located on the edge to the center line of the transfer bar at different speeds of vertical rolls: (a) – the point on the «head» part of slab (A); (b) – the point in the middle of slab's length (B); (c) – the point on the «tail» part of slab (C).

When assessing the minimum transition to the center line of the transfer bar, point C is taken (the point on the «tail» part of slab). Its transition is minimal in the mode when the vertical rolls create a support for horizontal rolls. For the remaining points (A and B), an inverse relationship is observed. When tension between vertical and horizontal rolls is created – the smallest transition is observed.

Such a multidirectional effect of the speed mode of vertical rolls is explained by various deformation conditions for the points under consideration. Points A and B were deformed in the horizontal rolls under the following conditions: the transfer bar was simultaneously in the vertical and horizontal rolls, which means tension or support takes place (depending on the mode). Point C was deformed in horizontal rolls under the following conditions: the transfer bar had already left the gap of vertical rolls.

Thus, the preferred mode of width reduction, from the point of view of minimizing the transition of metal from the sides and edges of the slab to the center line of the transfer bar, is a mode in which the absolute width reduction in all stands is uniformly distributed and linear speed of vertical rolls is lower than the linear speed of horizontal rolls.

IOP Conf. Series: Materials Science and Engineering **718** (2020) 012014 doi:10.1088/1757-899X/718/1/012014

4. Conclusion

Edge defects are the cause of jams, rejects and line downtime. Reducing the transition of the metal, as well as the defects, from the sides of the slab to the upper and lower surfaces of the transfer bar allows removal of the defective areas during subsequent trimming of the side edges. The effect of the distribution of width reduction of the slab and the speed of the vertical rolls on the amount of metal transition from the edges of the slab to the center line of the transfer bar is determined. An optimal mode of adjusting the operation of vertical rolls is proposed in order to reduce the amount of metal transition to the center line of the transfer bar.

References

[1] Ogarkov N N, Platov S I and Urtsev V N 2018 *Rolled products* **3** 15–21
[2] Xianghua L, Hailiang Y and Changsheng L 2006 *Proc. 9th Int. Steel Rolling Conf.* (France: Paris)
[3] Shabalov I P 2004 *Rolled products* **9** 3–12
[4] Chun M, Kwon H and Park H 2006 *Proc. 9th Int. Steel Rolling Conf.* (France: Paris)
[5] Konovalov Yu V, Ostapenko A L and Ponomarev V I *Calculation of sheet rolling parameters* (Moscow: Metallurgy) p 429

Algorithm and Design Methodology for Energy-Efficient Sheet Products Production Technology

D L Shalaevskiy[1] and A V Kozhevnikov[1]

[1] Cherepovets State University, 5 Lunacharsky street, Cherepovets, 162600, Russia

E-mail: shal-dmitrij@yandex.ru

Abstract. The work presents a calculation method for technology parameters of the continuous cold rolling process of wide steel strips, based on the determination of such reductions by stands and interstand tension, at which the technological process will be ensured with maximum productivity and minimum energy consumption. A method for calculating the rolling parameters is proposed, which will provide the minimum level of energy consumption by the main drives' motors of the rolling stands at the maximum productivity of the continuous mill by eliminating vibrations in the working stands. The created methodology can be used as a base for the software of continuous mills.

1. Introduction

Nowadays a sufficient number of methods for calculating the parameters of continuous cold rolling are known and applied [1-6]. The most widely used methods are based on indicators of the strength of the rolls, the permissible load on the drive elements of the working stands and the possibility of unifying the reduction and tension modes in some mill stands when rolling different profile and grades of the assortment, and some other methods. Experience in calculating rolling modes has shown that it is also advisable to use a calculation methodology that will enable the production of steel strips with the lowest energy consumption and the highest productivity.

Different sources already mention [7-9] the application of the methodology for designing technological parameters of the cold rolling process to ensure the minimum energy consumption and optimal performance by adjusting modes to exclude vibrations in the working stands. Positive experience has been accumulated in the search for optimal modes using well-known models of power parameters. However, in most studies, optimization methods known from the theory of experimental design were used to search for the optimal values of technology factors, which can lead to miscalculations and not full use of economical and increased productivity benefits. Obviously, to find the optimal values of the continuous rolling technology factors, the most rational approach is to compile a matrix containing combinations of all possible values of the technology parameters for a particular mill – local reductions in stands, interstand specific strip tensions and the corresponding total power of the main drive motors, determined using the most reliable methodology. From the matrix compiled this way, the most optimal option is selected according to the criteria “total energy consumption for the process” and “the absence of vibrations in the stands at maximum rolling speeds”.

IOP Conf. Series: Materials Science and Engineering **718** (2020) 012015 doi:10.1088/1757-899X/718/1/012015

2. Problem statement and research methodology

The objective of this work was to develop a method for designing cold rolling technology that provides the calculation of the least energy-consuming modes of strip rolling at maximum productivity of a continuous mill.

To conduct the study, we acted as follows. Determined the mechanical characteristics of the strip material - steel grade 08ПС - from the reference data. We selected a profile product range of rolled products - three conventional thickness groups: a thin strip, with a thickness of not more than 0.5 mm, a strip of medium thickness - not more than 0.8 mm, a thick strip, with a thickness of more than 0.8 mm.

Next, we compiled a table of technological parameters for each profile group of the range. The table included the reduction and the specific tension of the strip, found by looking through all possible combinations of these parameters, taking into account the technological limitations of the stands. An example of such a table is given below.

Table 1. Possible technological parameters of cold rolling on a 5-stand mill of a strip with a dimensions of 0.3 × 1020 mm (08Ю steel grade).

№	h_0, mm	h_1, mm	h_2, mm	h_3, mm	h_4, mm	h_5, mm	σ_1, MPa	σ_2, MPa	σ_3, MPa	σ_4, MPa	N_{dr}, MW
1	1.50	1.13	0.84	0.65	0.31	0.30	0.00	0.00	0.00	54.81	11.18
2	1.50	1.13	0.84	0.65	0.31	0.30	40.30	46.31	49.69	64.81	9.97
3	1.50	1.13	0.84	0.65	0.31	0.30	40.30	46.31	49.69	54.81	10.08
4	1.50	1.13	0.84	0.65	0.31	0.30	40.30	46.31	59.69	64.81	9.74
5	1.50	1.13	0.84	0.65	0.31	0.30	40.30	46.31	59.69	54.81	9.82
6	1.50	1.13	0.84	0.65	0.31	0.30	50.30	46.31	49.69	64.81	9.95
7	1.50	1.13	0.84	0.65	0.31	0.30	50.30	46.31	49.69	54.81	10.05
8	1.50	1.13	0.84	0.65	0.31	0.30	50.30	46.31	59.69	64.81	9.72
9	1.50	1.13	0.84	0.61	0.31	0.30	50.30	46.31	59.69	54.81	9.97
10	1.50	1.13	0.84	0.61	0.31	0.30	40.30	46.31	50.37	64.81	10.03
11	1.50	1.13	0.84	0.61	0.31	0.30	40.30	46.31	50.37	54.81	10.14
12	1.50	1.13	0.84	0.61	0.31	0.30	50.30	46.31	50.37	64.81	10.01
13	1.50	1.13	0.84	0.57	0.31	0.30	50.30	46.31	50.37	54.81	10.16
14	1.50	1.13	0.84	0.57	0.31	0.30	40.30	46.31	51.03	64.81	10.07
15	1.50	1.13	0.84	0.57	0.31	0.30	40.30	46.31	51.03	54.81	10.18
…	…	…	…	…	…	…	…	…	…	…	…
176	**1.50**	**0.90**	**0.83**	**0.76**	**0.31**	**0.30**	**45.20**	**55.30**	**61.66**	**64.81**	**9.24**
…	…	…	…	…	…	…	…	…	…	…	…
1962	1.50	0.82	0.76	0.69	0.31	0.30	46.60	57.20	61.17	0	10.02

Note: h_0 – the thickness of semi-rolled product,
$h_1 \ldots h_5$ – the thickness of the strip at the exit of 1-5 stands accordingly,
$\sigma_1 \ldots \sigma_4$ – specific tension at the exit from stands 1 to 4 accordingly.

For all the mode options created this way, the rolling forces and motor powers of the main drives were determined. To calculate the power parameters, a method was used that incorporates an elastic-plastic model of the deformation zone. The reliability of this mode is confirmed by publications [9].

Based on the study results of the obtained modes' parameters, we established the range of changes in the required motor powers of the main drives of the working stands for conducting the process on a rolling mill of a particular type. For a 1700 5-stand mill, when rolling a strip of 08Ю steel, 0.3 mm

IOP Conf. Series: Materials Science and Engineering **718** (2020) 012015 doi:10.1088/1757-899X/718/1/012015

thick and 1020 mm wide from a 1.5 mm semi-finished rolled product, this range was from 9.24 MW to 11.18 MW at a rolling speed of 12 m/s, i.e. the change in the parameter being optimized amounted to 17.3% of the minimum possible required power when changing the values of the technology factors.

Of all the combinations of parameters for each grade's group and profile assortment, one was selected that provided the minimum energy consumption for the process, taking into account limitations on the allowable rolling forces and motors power of the main drives. Such an optimal set of parameters is presented in table 1 in a row number 176. When we were looking for an optimal set of parameters, the total amount of processed data for each brand and profile assortment was about 2000 units. The optimal rolling mode was tested by the criterion of vibration resistance at maximum rolling speeds of up to 20 m/s.

The algorithm, which was used in order to determine the least energy-consuming mode of strip production is shown in Figure 1.

Studies using the algorithm were performed both for the stationary mode and for the mode in which the fluctuation of the strip thickness along its length was taken into account.

3. The results and its discussion

Modeling of power parameters for the modes calculated using the algorithm shown in Figure 1 made it possible to determine the amount of energy consumption at the existing mill and the potential for reducing energy consumption using the methodology proposed in [1].

The results of the assessment of energy consumption for the continuous rolling process at the 5-stand mill 1700 are presented in Figure 2.

The diagrams in Figure 2 allow us to conclude that there is a potential possibility of reducing process energy consumption. The largest number of possible technological modes - is up to 30% - that will ensure the energy consumption in the range of 8.9-9.2 MW at a rolling speed of 12 m/s. At the same time, the diagram shows that it is possible to conduct a process with a total energy consumption of less than 8.5 MW at the same rolling speed. There are less than 5% of such modes in the amount of all possible ones. It is advisable to reduce the energy consumption to conduct the rolling process in accordance with the technology that provides this minimum level of energy consumption. It is established that the process with the specified energy consumption ranges at speeds of 12 m/s is possible without the occurrence of vibrations in the stands of a continuous mill.

The results of modeling the rolling modes produced on a significant amount of data made it possible to establish the influence of technology parameters on the level of energy consumption. Thus, the change in the total required power for conducting the rolling process at the 5-stand mill, depending on the local reductions in the stands and the specific tension is shown in Figure 3.

From the dependencies shown in Figure 3, it can be seen that the highest decrease in the total energy consumption is influenced by the values of the local reductions in the first and second-to-last stands, as well as the values of the interstand tensions. Moreover, the increase of reduction in the first and second-to-last stands and all the specific tensions reduce the level of process energy consumption.

These studies of technology parameters finally made it possible to establish the nature of their influence on the amount of energy consumption:

- a decrease in the total required power for the entire studied profile assortment is ensured by an increase in local reduction in the first stand, and this value has the most significant effect when rolling the thicker strips;
- the increase of specific tensions in all interstand gaps will have an additional impact on the decrease of required power when producing strips of all steel grades regardless of the work roll barrel diameter, the decrease of the total energy consumption with the increase of the specific tension in one of the interstand gaps on 10MPa may reach the value of 120 kW with the rolling speed up to 12 m/sec;
- unloading of the intermediate stands by the value of local reduction with the transition of the remaining reduction to the first and second-to-last stands ensures a reduction in the required total drive power of the working stands in the production of strips with a thickness of less than 0.5 mm;

IOP Conf. Series: Materials Science and Engineering **718** (2020) 012015 doi:10.1088/1757-899X/718/1/012015

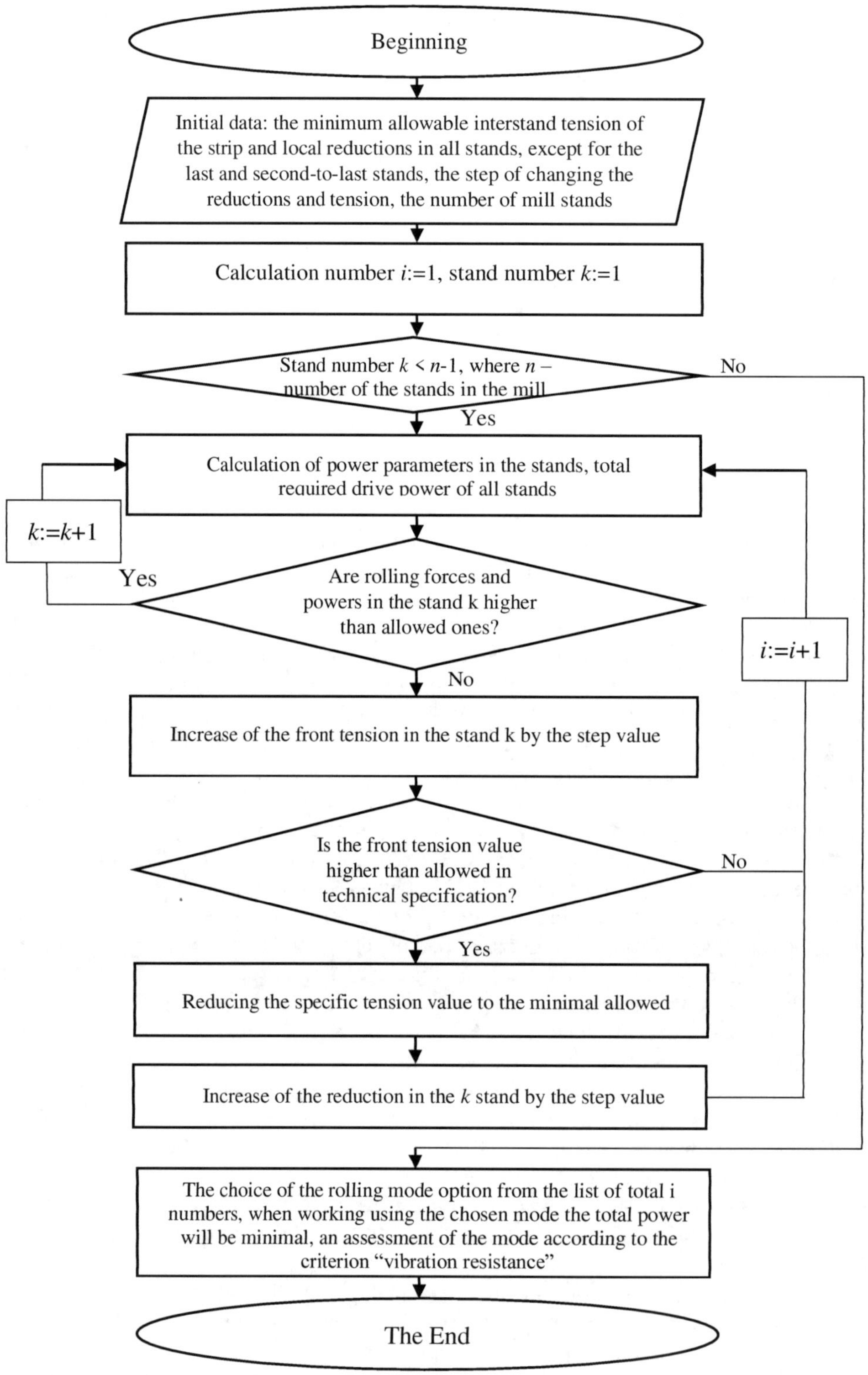

Figure 1. Algorithm for determination of energy-efficient mode.

IOP Conf. Series: Materials Science and Engineering **718** (2020) 012015 doi:10.1088/1757-899X/718/1/012015

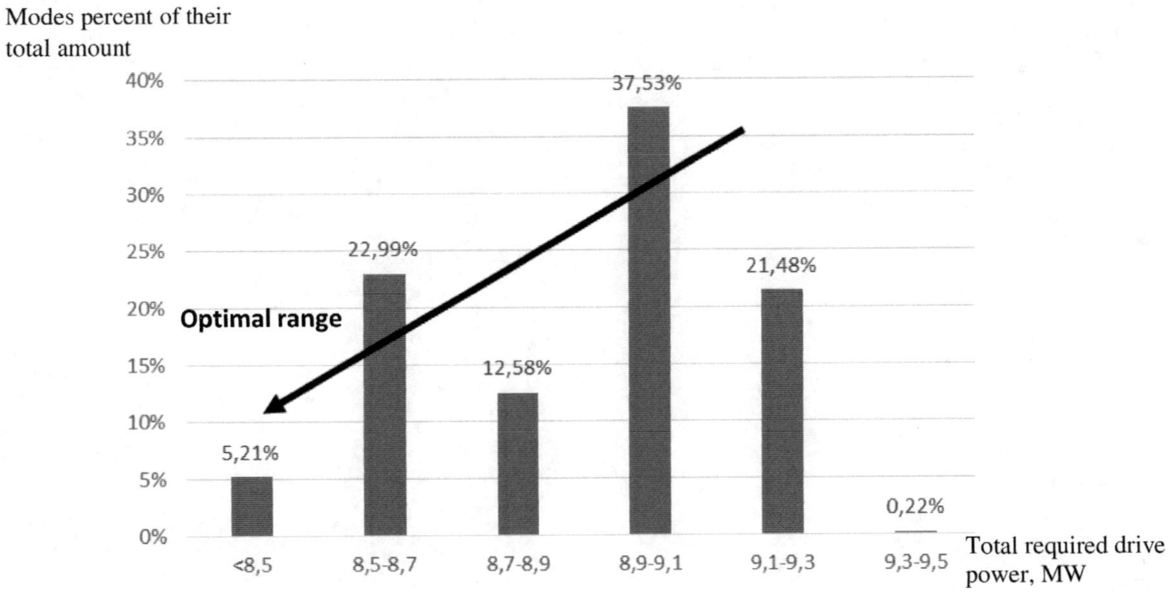

Figure 2 – Distribution of total energy consumption when using all possible combinations of technology parameters for rolling a strip of 0.5 mm thick (08ПС steel) with a speed of 12 m/s.

- when the reduction of the strip with thickness more than 0.5 mm takes place at the second-to-last stand, the increase of the energy discharge can be observed, but when rolling thinner strips, the opposite – energy discharge reduces, while the reduction increases by 2%, the change of energy consumption may reach the value of 150 kW with the rolling speed of 12 m/sec.

The least energy-consuming rolling modes were also analyzed by the values of technological parameters, the results of this analysis for a 5-stand mill are presented in Figure 4 for finished strips with a thickness of 0.3 to 0.5 mm.

It was also established that the proposed algorithm and methodology will be most effective when designing the processing modes of the thinnest strips, as can be seen from the diagram in Figure 5.

According to the results of the study, it was determined that the diameter of the roll barrel and the steel grade of the strip will not affect the efficiency of the methodology.

As an example of using the created methodology in [1], presented is the rolling mode for a 5-stand continuous mill with a barrel diameter of the work roll in all stands equal to 600 mm, which ensures a reduction in the required power for the process by 3.5% compared with the current mode used on a 5-stand mill 1700.

It should be noted that due to thickness fluctuations along the length of the semi-finished rolled product, there are also fluctuations in the required drive power in the working stands. Thus, when changing the thickness of the strip in front of the first stand by 5% of the nominal value, the level of power fluctuations can increase up to 3-4% of its average value when the mill is working (for example, 1700 5-stand mill) using the existing technology.

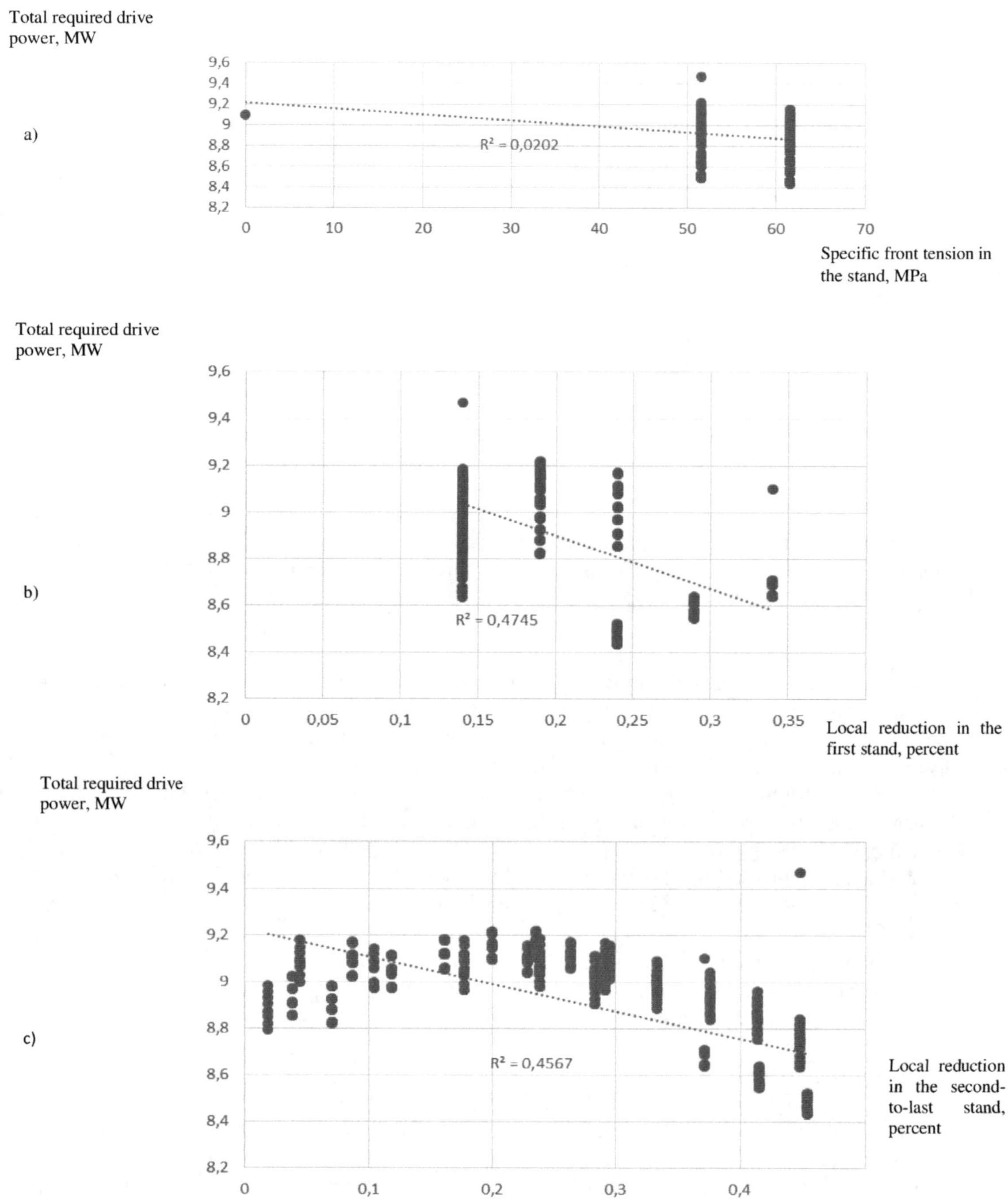

Figure 3 – The influence of the specific tension in the last interstand gap on the total required drives' motors power of the continuous mill stands a) local reductions in the first b) and second-to-last c) stands

IOP Conf. Series: Materials Science and Engineering 718 (2020) 012015 doi:10.1088/1757-899X/718/1/012015

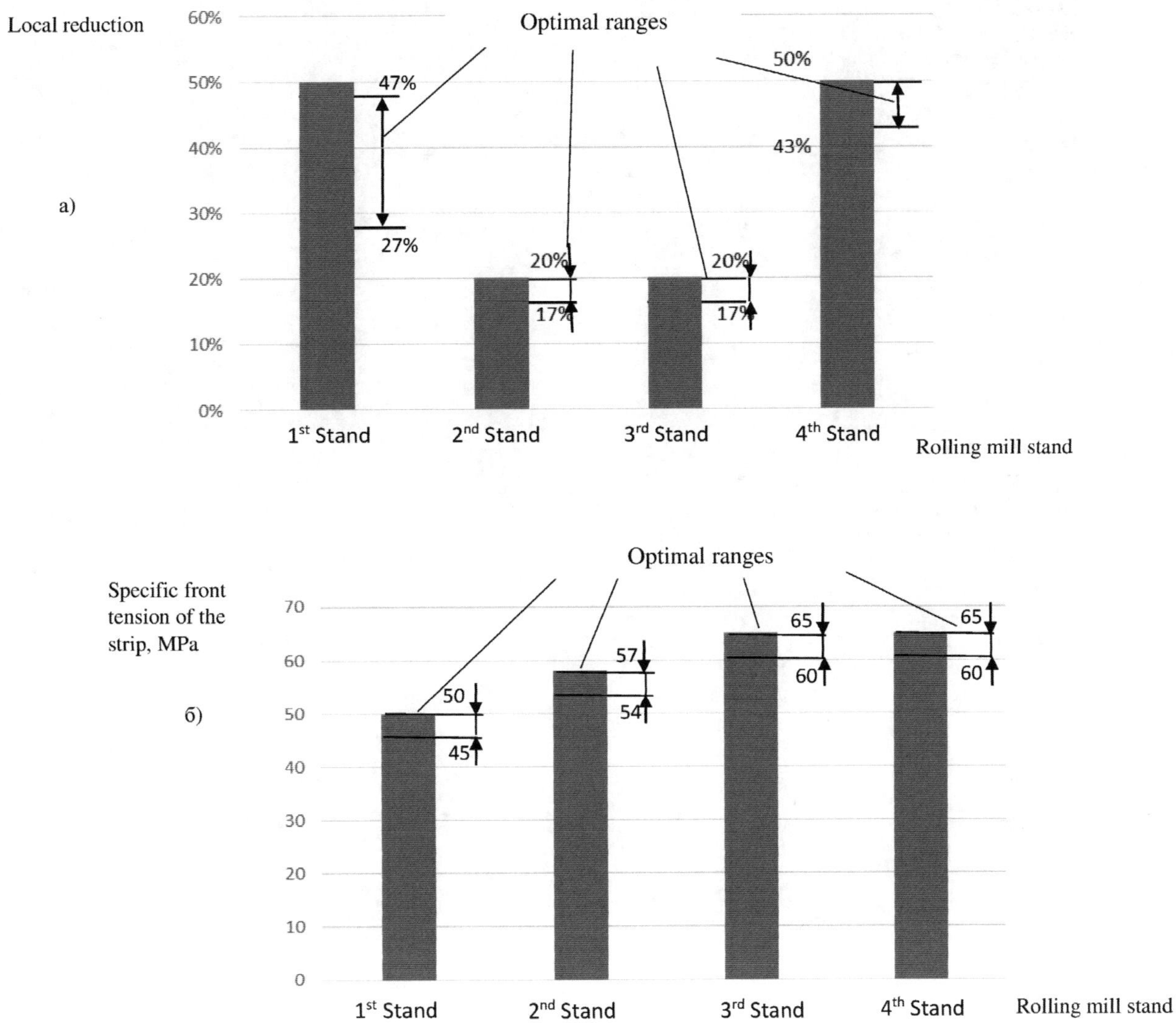

Figure 4 – Ranges of the local reductions a) and specific tensions b) in work stands of the 5-stand mill 1700, providing minimal energy consumption when rolling

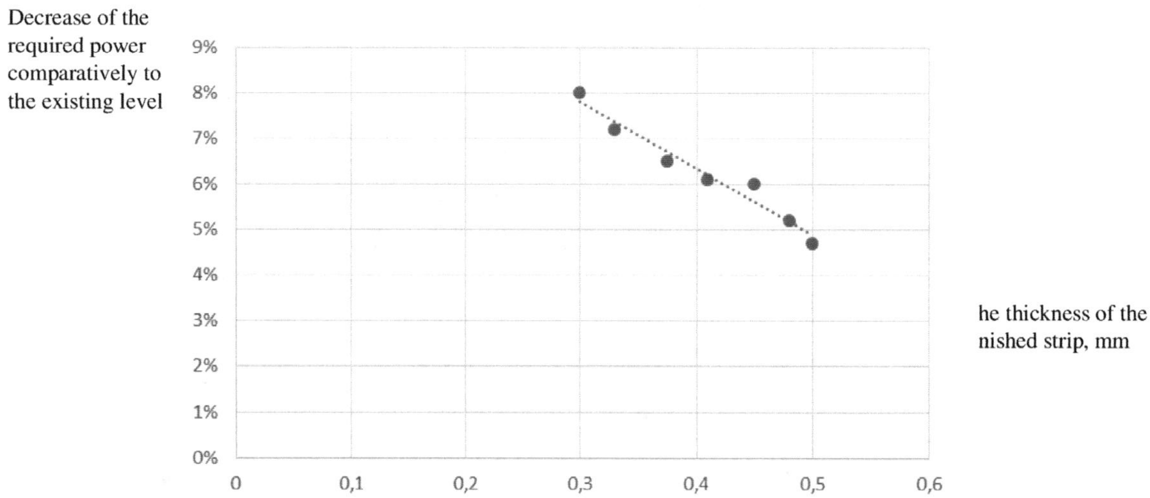

Figure 5 - The diagram that shows the influence of the thickness of the finished strip on the decrease of the required power of rolling process

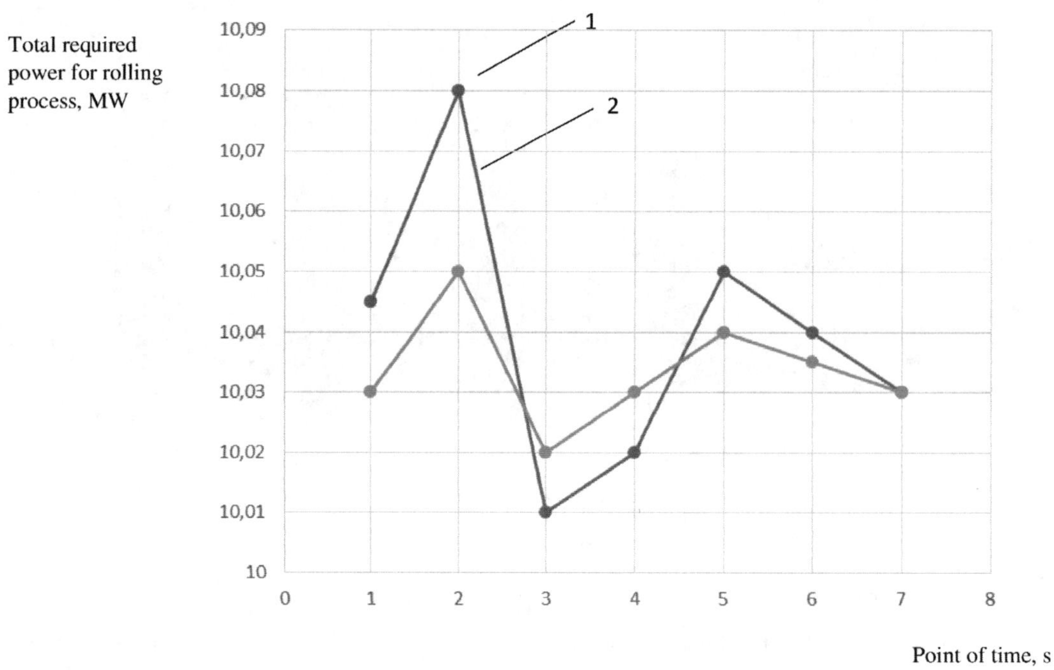

Figure 6 - The diagram that shows changes in the required power of the main drive motors of the stands (over time) when rolling a strip 0.5 mm thick made of 01ЮТ steel when operating a 5-stand mill 1700 according to existing technology (1) and developed using the newly proposed automated algorithm (2)

The experience of applying the newly proposed algorithm showed that energy-efficient modes provide a reduction in power fluctuations by 1-2%. A fragment of the diagram that shows changes in the momentary required power (over time) for the rolling process is shown in Figure 6.

The decrease in power fluctuations when using energy-efficient modes can be explained by the fact that at each moment of time, when the thickness of the semi-finished rolled product changes, the rolling technology will be most optimal from the point of view of energy consumption.

The results of computer-aided design of modes for a wide range of assortments made it possible to formulate general recommendations on the selection of technology parameters (see Table 2). It is worth noting that when implementing these recommendations, it is necessary to take into account the limitations of the working stands in terms of rolling forces and power, as well as the conditions for ensuring a stable rolling process.

Table 2. Recommendations for choosing the parameters of the reduction and tension of the steel strip during continuous rolling.

Parameter	Value
Specific interstand tension	The maximum possible of the regulated values is 22% of the yield strength of the strip material.
Reduction in the first stand	Reduction, that provides the rolling force or the first stand main drive power loading for 65-75 % of the maximal allowed values.
Reduction in second-to-last stand	Reduction, that provides the rolling force or the first stand main drive power loading for 65-75 % of the maximal allowed values, when producing strips with the thickness less than 0,5 mm.
Reduction in the last stand	4-8 % when rolling in the 5-stand continuous mill, 8-12% when rolling in the 4- stand mill
Reductions in intermediate stands	Equal distribution of the remaining total reduction between the intermediate stands, according to the condition of equal rolling force distribution in these stands.

Using algorithm for checking the technological mode for vibration resistance can improve the performance of a continuous mill during the process.

The results of the work can be used in automated control systems for the technological process of continuous mills when determining the modes of reduction and tension of the steel strip.

4. Conclusion

This work presents parameters' calculation method for the continuous cold rolling technology of wide steel strips. The method is based on the definition of such reductions in every stand and interstand tensions, that will provide the running of the technological process with maximum performance and minimum energy consumption. The type of the steel strip reductions and tensions that affects the amount of energy consumption and operating is determined, as well as the effect of different steel profiles and grades, diameter of the work rolls barrels in the stands. A way for calculation of rolling parameters is suggested, that will provide the minimal level of the power being consumed by the main rolling stands drives' motors with maximum performance of the continuous mill by means of excluding the vibrations in the work stands. The created method can be used in the continuous mills software.

References

[1] A. Kozhevnikov, I. Kozhevnikova, N. Bolobanova, D. Shalaevskiy, Methodology of design of thin steel strip cold rolling energy efficient technology, Metalurgija (Zagreb, Croatia), 2018, T. 57, № 4, P. 337-340.

[2] Storck J. Enabling Manufacturing Competitiveness and Economic Sustainability Product variety, flexibility and energy use in hot rolling mills. Springer, 2011. P. 80-85.

[3] Shilov V. A., Ponomarev A. A. Optimization of technological rolling modes according to the criteria of mechanical properties, Production of flat-rolled products. 2013. № 2. P. 14-19.

[4] Genkin A. L. Энергосберегающее управление листопрокатным комплексом “печи-стан”// Steel. 2011. № 3. P. 34-40

[5] Rumyantsev M. I. Some results on the development and application of a methodology for improving sheet-rolling technological systems, Bulletin of Magnitogorsk State Technical University named after G. I. Nosov, 2017, T. 15, № 1, P. 45-55.

[6] Rumyantsev M. I. Methods for the development and improvement of technology for the production of rolled metal to improve sheet rolling systems, Theory and technology of metallurgical production. 2017, № 4 (23), P. 26-36.

[7] Garber E. A., Shalaevsky D. L., Mitrofanov A. V., Mishnev P. A., Ogoltsov A. S. Development and industrial research of energy-efficient hot rolling modes for wide steel strips, Ferrous metallurgy. 2016, № 5, P. 38-43.

[8] Kuznetsov V. V., Garber E. A., Yusupov V. S., Shalaevsky D. L. Technological features of the production of cold-rolled automotive sheet from new corrosion-resistant sparingly alloyed steel with BH-effect, Production of flat-rolled products. 2007, № 11, P. 9-12.

[9] Garber E. A., Shadrunova I. A., Traino A. I. Determination of power parameters for cold rolling of the thinnest strips, Ferrous metallurgy. 2002, № 2, P. 47.

[10] I.A. Kozhevnikova, N.L. Bolobanova, A.V. Kozhevnikov Cold rolling with vibration of the working rollers // Steel in Translation, 2017, Vol. 47, No. 10, pp. 695–698.

[11] KozhevnikovA. V., Bolobanova N. L., Kozhevnikova I. A. The study of influence of technology factors and equipment design parameters on energy consumption taking into account stochasticity of the cold rolling process // 2017 IOP Conference Series: Materials Science and Engineering.

Acknowledgements

Financial support was provided by the Russian Ministry of Education and Science (project no. 11.3943.2017/4.6).

IOP Conf. Series: Materials Science and Engineering **718** (2020) 012016 doi:10.1088/1757-899X/718/1/012016

Formation of the Structure and Decarburized Layer During Hot Screw Rolling of Vessels

R L Shatalov[1] **and V A Medvedev**[2]

[1] Moscow Polytechnic University, 38 Bolshaya Semenovskaya street, Moscow, 107023, Russia
[2] Bogorodsk branch of JSC "SPA "Pribor", 1 Kirovogradskaya street, Moscow, 117519, Russia

E-mail: mmomd@mail.ru

Abstract: We studied the influence of temperature and soaking time on rod billets made of steel 50. The development of the decarburized layer after induction heating and soaking in the resistance furnace was taken into account. When analyzing cooled down rod billets subjected to heat treatment, a continuous and stable development of a ferritic layer on the entire surface of rod billets was detected. Rod billets heated in a similar way as those on which the size of the outer decarburized layer had been defined were hot rolled. Experimental rolling was performed in a 3-high stand of the rolling-press line of JSC Scientific Production Association (SPA) "Pribor". After cooling it was found that the sizes of a ferrite ring on the outer surface of steel vessels had changed. The analysis of the obtained data allowed us to find a relationship between the change in the structure of a ferrite ring formation and a draw ratio λ. A regression equation was obtained for determining the influence of drawing during screw rolling on the size of a decarburized layer.

1. Introduction

Recently, in metallurgy and a machine-building industry rolling-press lines (RPL) have been used for producing hollow vessels with a diameter of 30–60 mm using carbon structural steel. These lines allow to produce the items of the required shapes and sizes from round rod billets by consecutive hot screw rolling and stamping [1-3].

The purpose of the vessel production on RPL of JSC SPA "Pribor" is to reduce the expenses by increasing the material utilization rate (MUR) and reducing the machining time. Cut-to-length rod billets made of steel 50 and produced according to GOST 1050-88 with the diameter of 42 mm at small-section mills of Nadezhda Metallurgical Plant are used as a basic material.

In order to increase the ductility of steel 50, rod billets are heated in an induction furnace and soaked in a resistance furnace, which results in the decrease of carbon content and the formation of a decarburized layer on metal surface which remains during further deformation [4].

During hot rolling the rod billets made of steel 50 change their size and shape, at the same time the concentration of ferrite on the external surface of the products changes and a so-called ferrite ring is formed. A decrease in carbon concentration in steel results in the increase of plastic properties and the decrease in strength characteristics [5, 6].

IOP Conf. Series: Materials Science and Engineering **718** (2020) 012016 doi:10.1088/1757-899X/718/1/012016

The purpose of the research is to determine the influence of hot deformation of rod billets on the change of a surface decarburized layer during screw rolling and to relate the drawing to the change in size of the ferrite ring.

2. Materials and experiment methods

As a basic material we used hot rolled machined rod billets with the length of 70 mm and the diameter of 42 mm made of steel 50 and produced at Nadezhda Metallurgical Plant (NMP) according to GOST 1050-88, heat No. 7877, batch 061721363 with the specified mechanical properties: tensile strength – 856 MPa; yield strength – 660 MPa; relative narrowing – 43%.

The size of the decarburized layer on the surface of initial rod billets supplied by NMP was studied with CarlZelss optical microscope following the methods of microstructure study GOST 5640-68. Figure 1 shows the microstructure taken from a cylindrical surface of the rod billets made of steel 50 and supplied by NMP at a100x magnification.

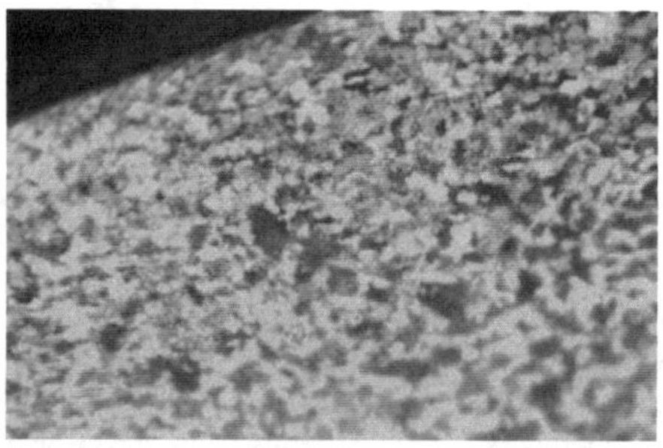

Figure 1. Microstructure from a cylindrical surface of the rod billets made of steel 50 and supplied by NMP shown at a100x magnification.

Microstructure shown in Figure 1 has a homogeneous distribution of ferrite and perlite in the entire studied field at the ratio of 2:1, the grain size is approximately 7-8 scores. No traces of decarburized layer ferrite were found on the surface of rod billets.

Rod billets made of steel 50 were preheated in the induction furnace to 1160 °C for 12 seconds.

The equalizing soaking of rod billets took place in the resistance furnace and continued for 20 minutes at the temperature of 1160 ± 10°C.

The size of the surface decarburized layer was studied following the method of microstructure study GOST 5640-68 for the rod billets made of steel 50 which were subject to heat treatment at 1160 °C with a total time equal to the heating and soaking cycles, while cooling took place in oil.

Figure 2 shows the microstructure taken from the cylindrical surface of rod billets made of steel 50 and subject to heat treatment. The studies were carried out at a 100x magnification.

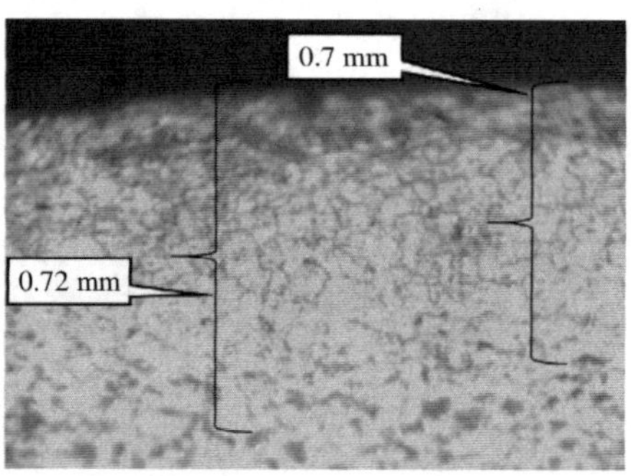

Figure 2. Microstructure from the cylindrical surface of rod billets made of steel 50 after a heat treatment shown at a 100x magnification.

As it can be seen in Figure 2, the ferrite layer on the surface of rod billets made of steel 50 is continuous. The deviations of ferrite content on the studied cylindrical surface are within the range from 0.7 to 0.72 mm, which is no more than 3%.

During the rolling of rod billets made of steel 50 which had been heated and soaked at the temperature of 1160 °C in "30-80" RPL mill of JSC SPA "Pribor" plant the relative deformation and draw ratios λ were checked [7]. After screw rolling the relative deformation of the top part ε_t and the bottom part ε_b, as well as the spread of average draw ratios λ were calculated.

The relative deformation of the vessel bottom

$$\varepsilon_b = ((F_0 - F_b) / F_0)\cdot 100\%. \quad (1)$$

The relative deformation of the vessel top

$$\varepsilon_t = ((F_0 - F_t) / F_0)\cdot 100\%. \quad (2)$$

where the cross-section area in both cases $F_0 = \pi/4\cdot D_0$.

The draw ratio was defined

$$\lambda = L_i / L_f. \quad (3)$$

where L_i is the initial length of a rod billet and L_f is the final length of a rod billet after rolling, respectively.

3. Discussion of the research findings

During the processing in a 3-high stand of "30-80" screw rolling mill with a further placement on a rotating mandrel the dimensions changed: outer diameter was 36 ± 0.5 mm, inner diameter was 18 ± 0.5 mm, total length was 120 ± 1 mm and cavity depth was 105 ± 1mm. After rolling with a cavity formation the relative deformation of the steel vessel bottom ε_b was about 26% and ε_t of the top part was about 45%.

The spread of the drawing values after rolling was λ = 1.70-1.72.

We studied the size of the decarburized layer after rolling in "30-80" line for the vessels made of steel 50 after cooling in oil.

Figure 3 shows the microstructure taken from the cylindrical surface of a vessel made of steel 50 which was subject to hot rolling. The research was done at a 100x magnification.

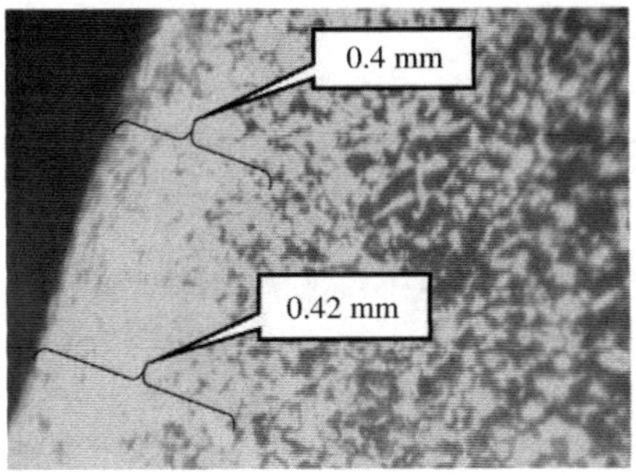

Figure 3. Microstructure of a vessel made of steel 50 which was subject to hot rolling shown at a 100x magnification.

As it can be seen in Figure 3, ferrite is distributed on the entire cylindrical surface of the vessel made of steel 50 uniformly. No traces of the ferrite ring discontinuity were found. The size of ferrite on the surface decreased as compared to the original rod billets (see Figure 2) and is within the range from 0.4 to 0.42 mm, which is no more than 5% of the limit values. There is a diffusional decay structure underneath the layer of pure ferrite of a decarburized ring resulting from incomplete quenching in oil – sorbitol with the traces of ferrite inclusions at the ratio of 2:1, respectively.

The research findings helped to determine the regularity of the drawing effect after finishing rolling of the vessels made of steel 50 and produced in RPL line of JSC SPA "Pribor". A consistent relation between the drawing after rolling and the change in ferrite ring sizes for the outer part of a vessel was obtained. Figure 4 shows the influence of drawing during rolling on the thinning of the surface ferrite layer.

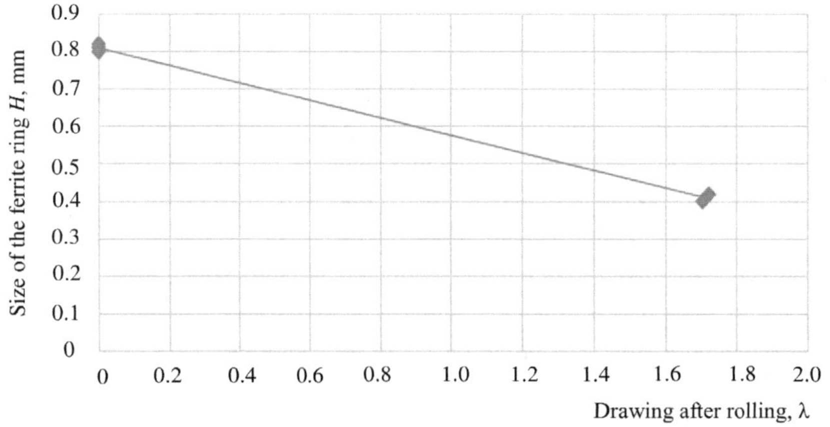

Figure 4. The dependence of the change in the ferrite ring size on the drawing deformation.

The research findings allowed us to derive a regression equation using the methodology from the research [8] that allows to calculate the change in the outer ferrite layer during the elongation of a rough rolled vessel made of steel 50 after screw rolling:

$$H = 0.7995 - 0.2477\,\lambda. \qquad (4)$$

The obtained equation allows to evaluate the change in the surface decarburized layer of cylindrical billets made of steel 50, as well as during rolling, with a traditional and modern method. The statistical model has a high correlation factor R^2, more than 0.94, which indicates a high reliability of the established relation.

4. Conclusions

1. The influence of heating temperature and soaking time on the thickness of the decarburized layer of rod billets made of steel 50 is shown. It has been found that a continuous ferrite layer is within the range of 0.70-0.72 mm.
2. Rod billets made of steel 50 were rolled in RPL rolling mill. The deformation of the rolled vessel areas was calculated: in the bottom area ε_b is about 26% and in the top area it is about 45%. The spread of drawing values after rolling was determined as $\lambda = 1.70\text{-}1.72$.
3. It has been found that after screw rolling with the drawing of 1.70-1.72 the structure and the sizes of ferrite near a cylindrical surface of a steel vessel change. The thickness of a decarburized layer ranges from 0.4 to 0.42 mm.
4. The laws of the influence of a deformation degree and drawing on the change in thickness of a surface decarburized layer of vessels made of steel 50 have been established.
5. A regression equation that links the drawing value and the change in size of the decarburized layer on the surface of vessels made of steel 50 during screw rolling was obtained. The obtained equation can be used in the design of rational deformation modes.

References

[1] Romanzev B A, Goncharuk A V, Aleshchenko A S et al 2015 *Metallurgist Journal* **5** 41–43
[2] Romanzev B A, Gamin Y V, Goncharuk A V and Aleshchenko A S 2017 *Metallurgist Journal* **3** 336–338
[3] Gamin Y V and Romanzev B A 2015 *Rolled Metal Production Journal* **11** 25–31
[4] Gorlova A A and Rodinkov S V 2008 *Ferrous Metals Journal* **1** 14–18
[5] Galkin S P, Romanzev B A, Suan Ta Din and Gamin Y V 2018 *Ferrous Metals Journal* **4** 2–6
[6] Filippov A A, Pachurin G V, Kuzmin N A and Matveev Y I 2018 *Ferrous Metals Journal* **4** 36–40
[7] Shatalov R L and Medvedev V A 2019 *Metallurgist Journal* **2** 53–58
[8] Shatalov R L, Lukash A S and Ziselman V L 2014 *Non-Ferrous Metals Journal* **5** 61–65

IOP Conf. Series: Materials Science and Engineering **718** (2020) 012017 doi:10.1088/1757-899X/718/1/012017

Aeroacoustics of Jet Streams Used in Metallurgy

O Shlik[1]

[1] Accusteel Ltd., Netanya 6/a4, 4240718, Israel

E-mail: shlikoleg@gmail.com

Abstract. The supersonic jets of gas which is expiring into the melt from nozzle of BOF lance have a powerful acoustic radiation. The sound intensity of the jets is up to 200 dB and it is the most powerful natural source of sound. The jets invasion into heterogeneous unstable and gas saturated liquid melt-solution is the cause of the cavitation phenomena into the liquid. This explains the spontaneity of the gas formation and the speed of the decarburization process throughout the entire volume of the liquid metal volume. Using sound intensity changes, in the 60-th years of the last century, was developed the acoustic monitoring and control system of slag process into the BOF furnace.

In the 21st century, the metallurgical industry is going through more computerization of control process of the technological process of steelmaking. The new knowledge about the mechanism of decarburization processes can provide solution to the automation problem.

Several hypotheses of the decarburization process of steel in blast oxygen furnace (BOF) are known. The first hypothesis is one-stage, where the decarburization process occurs in the zone of contact of the oxidizing jet with the iron-carbon melt solution. High temperatures in the zone of interaction of the oxidizing agent with the melt facilitate to the diffusion of carbon from the entire volume of the melt into the high-temperature reaction zone. The second hypothesis is two-stage, where iron oxides formed in the reaction zone of an oxygen jet with a melt-solution diffuse into the melt providing intensive boiling of the melt bath with the release of CO bubbles throughout [1]. There is no doubt that the existing hypotheses of both the one-stage and two-stage decarburization process take place and are more characteristic of the open-hearth process, where the decarburization process lasts for hours. But not one of the hypotheses of the mechanism of the decarburization process presented above is unable to explain the high speed of decarburization processes throughout the entire volume of the melt bath, which is typical for the BOF process.

Perhaps there is a hypothesis that allows us to explain the high speed of the decarburization process in the BOF process.

The work [2] shows the intensification of the decarburization of the melt by mechanical vibrations of the oxidizer jet with a frequency of 800 - 900 Hz. It presents a lot of mechanical and gas-dynamic devices that provide an intensification of the decarburization process. The process acceleration mechanism is not explained, but the fact of acceleration of the decarburization process is ascertained.

Regard the aero-acoustic properties of supersonic jets flowing into fulfilled space.

Figure 1 (a) shows supersonic shock wave part of the jet and subsonic blurred part with a small power of turbulence. As a result of the interaction of the jet turbulence with the space surrounding the jet, the supersonic section of the jet dissipates passing into the subsonic part of the stream. Figure 1 (b) shows the outflow from the same nozzle in the same mode when the jet was excited by its own acoustic radiation. The own acoustic radiation reflected from the acoustic screen mounted on the surface of the nozzle enter the jet into resonant self-excitation providing the feedback loop. Changing the shape or position of the screen changes the feedback efficiency [7], which leads to a change in self-oscillating processes in the jet with changing the size of the turbulence.

a

b

Figure 1. Shadow photograph of a supersonic jet flowing from lance nozzles into full filled space (exposure 10^{-6} s).

In the nearest acoustic field near the supersonic part of the jet the radiation intensity increases to 200 dB. The directivity of acoustic radiation determined by the scanning method [3] is shown in Figure 2. The acoustic field of the supersonic jet has a specific directivity, along the stream of the jet and against the stream. The radiation directed against the stream is formed by the turbulence radiation of the supersonic part of the jet and has a maximum intensity at the specific frequency. The radiation by direction of the jet's flow is formed by the supersonic and subsonic part of the jet and is a source of broadband white noise. As can be seen in the shadow photograph Figure 3 [4].

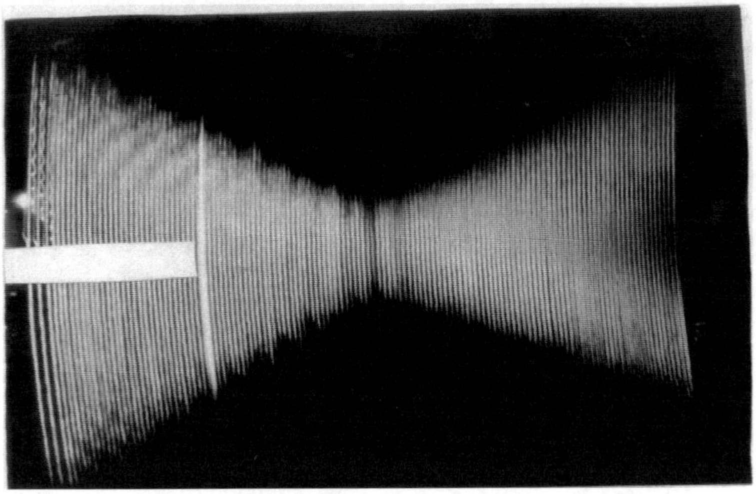

Figure 2. Acoustic radiation of the supersonic jet.

Figure 3 shows the shadow photograph of acoustic radiation of the supersonic jet. The generation mechanism is the turbulence of the shock-wave structure of a supersonic jet. The intensity of this radiation (120-200dB for supersonic) prevails over the turbulent acoustic radiation of the subsonic part (100-110dB). The supersonic jet is a natural source of acoustic radiation. The intensity and directivity of the radiation depends on the flow regime, size, shape and design of the nozzle block.

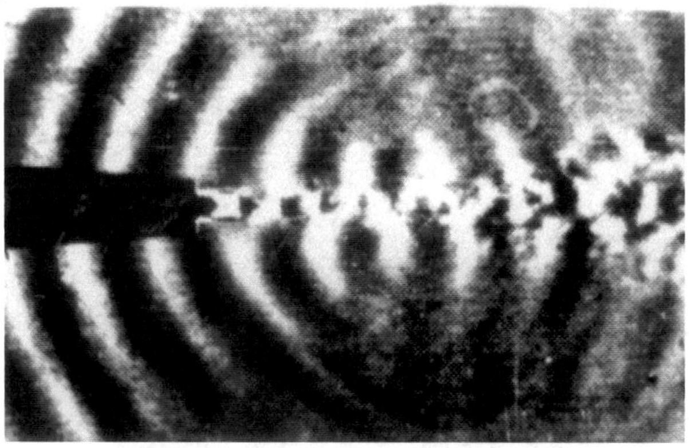

Figure 3. The shadow photo of the acoustic radiation of the supersonic jet flowing into full filled space.

An analysis of the operation of factory lances shows that the jets flowing from the nozzles have this radiation and have an effect on the technological process.

From the foregoing, a new hypothesis of the decarburization process can be formulated. The third hypothesis of the decarburization process of the melt-solution is wave model. The melt-solution in its physical nature presents 95% iron in which 4.3% carbon 0.07% of other chemical elements of the periodic table are dissolved. In addition to chemical elements, gases, including oxygen, are dissolved in the melt-solution.

In the process of introducing a supersonic jet of an oxidizing agent into the melt-solution [6–7], the torch of combustion of the oxidizing agent CO gas and metal is formed. With increasing temperature and intensity of acoustic radiation, the magnitude of the supersonic part of the jet increases. The sound wave of high intensity cause to a break in the continuity of the melt-solution medium at concentrators, which are carbon dissolved in the melt. The cavitation and collapse at the place of the bubbles is accompanied by an adiabatic local temperature of more than 10,000 ° C [5]. The process is spontaneous in nature throughout the entire volume of the melt-solution bath, causing to the CO gases formation. This explains the high speed of the decarburization process of the melt.

The evidence of the transfer of jet oscillations to the melt with the formation of wave phenomena in the melt is presented in Figure 4 [2].

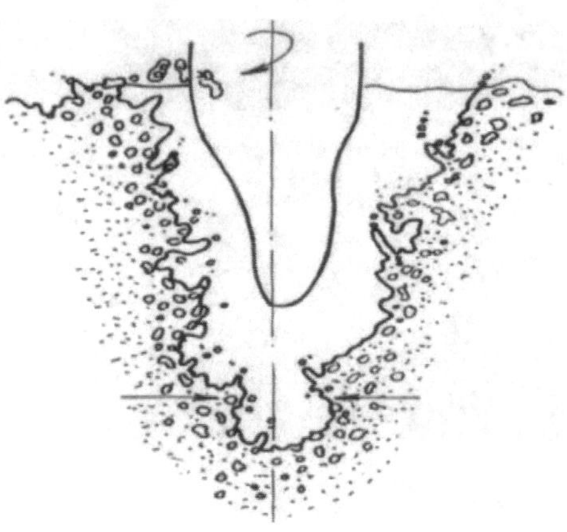

Figure 4. The reaction zone of the introduction of a supersonic jet into the melt.

Figure 4 [2] shows the surface of the reaction zone, where the vibrations of the supersonic part of jet are transferred into the melt. The cavitation is observed in the form of small bubbles throughout the reaction zone. The supersonic part of jet grows when it expires into the high-temperature space [6]. With an increase of the supersonic part of the jet, the intensity of acoustic radiation and the effect of acoustic feedback are increased [7]. The outflow of the supersonic jet of oxidizing agent into the melt is accompanied by the formation of a torch. The torch stoichiometry shows that the combustion temperatures are 3000-5000 ° C. The acoustic radiation penetrates unhindered into the melt. This is confirming the wave hypothesis of the decarburization process.

Acoustic radiation spreads in the cavity of the steelmaking unit and scans the gas medium, extracting information about the state of the melt. For the first time the acoustic phenomena of jets of blast was used for monitoring and control of the slag phase of the BOF process [8-9]. In the 60s of the last century, acoustic systems for monitoring and controlling the slag-metal phase of the BOF process were introduced (Figure 5).

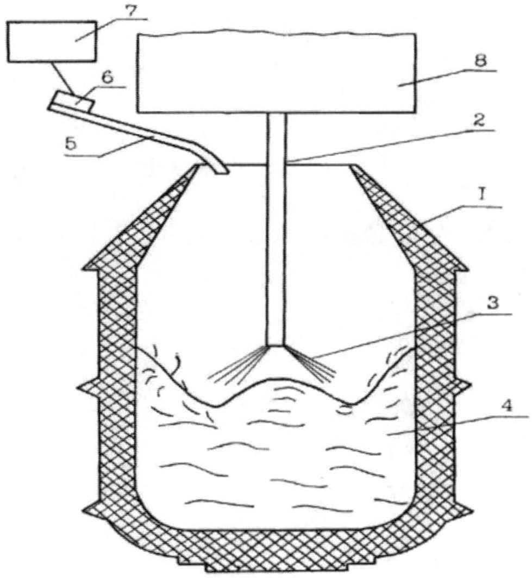

Figure 5. 60s acoustic system: 1 - blast oxygen furnace (BOF); 2 - oxygen lance; 3 - supersonic jets flowing from the lance nozzles into the melt - 4; 5 - waveguide; 6 - microphone; 7 - electronics of the acoustic system, which consists of band-pass filters, the receiver of the acoustic signal converted into electric.

The result of processing the acoustic noise signal is transmitted to the recorder, which fixes the received amplitude of the acoustic radiation signal to the chart tape in real time, Figure 6.

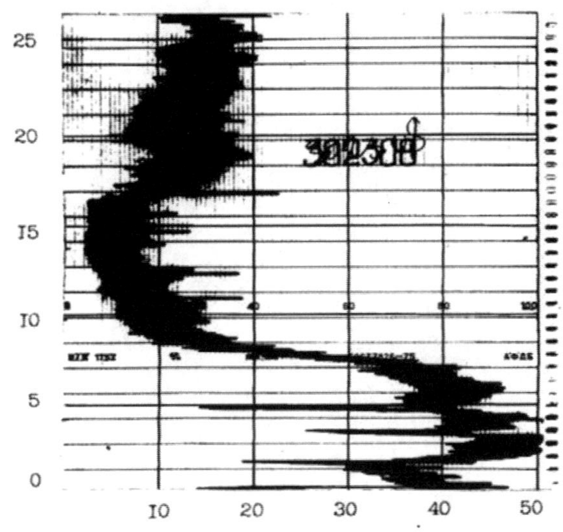

Figure 6. The slag behavior diagram during the smelting process.

IOP Conf. Series: Materials Science and Engineering **718** (2020) 012017 doi:10.1088/1757-899X/718/1/012017

The changes of amplitude recorded on the diagram correspond to visual observation of the behavior of slag-metal emulsion in the unit. The slag and blast modes of smelting were optimized using process information. The lack of knowledge about the features of aero-acoustic radiation from jets flowing from nozzles of lances, burners, dispersants causes to the experimental search for the optimal design of the device.

References

[1] Yavoisky V I 1966 *Theory of steelmaking processes* (Moscow: Metallurgy) p 792

[2] Yavoisky V I, Yavoisky A V and Sizov A M 1985 *The use of pulsating blast in steel production* (Moscow: Metallurgy) p 174

[3] Kok U E 1974 *Visible sound* (Moscow: Mir) p 120

[4] Davies M G and Oldfield D E S 1962 *Acustica* **12** 257–277

[5] Pernik A D 1968 *Cavitation problems* (Moscow: Nauka) p 166

[6] Glotov G F and Kuzemko R D 1975 *In didgest: Heat and mass transfer processes in bathtubs of steelmaking units. Publications of the I Vsesouznij Scientific Conference (*Moscow: Metallurgy) p 45–51

[7] Kapustin E A, Nescheret P A, Ryabukhin A V, Shlik O E and Voroshilin V.S 1982 *News of universities. Ferrous metallurgy* **11** 120–123.

[8] Turkenich D I 1971 *Management melting steel in the BOF* (Moscow: Metallurgy) p 310

[9] Turkenich D I and Zdanovskij V V 1978 *Acoustics in BOF smelting technology* (Moscow: Metallurgy) p 80

IOP Conf. Series: Materials Science and Engineering **718** (2020) 012018 doi:10.1088/1757-899X/718/1/012018

Nanomodification of Refractories with Finely-Dispersed Additives with the Use of a Vortex Electromagnetic Homogenizer

D S Suvorov[1], B B Khaidarov[1], D V Lysov[2], D V Kuznetsov[1], P A Zhukova[1], V I Kazakov[2] and V V Gorchakov[3]

[1] National University of Science and Technology MISIS, 4 Leninsky prospect, Moscow, 119049, Russia
[2] OOO Funktsionalnyie nanomaterialy (Functional nanomaterials), 4 Leninsky prospect, Moscow, 119049, Russia
[3] OOO NPP Intor (Research and Production Enterprise Intor), 39 Troitskaya Street, Novocherkassk, 346428, Russia

E-mail: suvorov8225@yandex.ru

Abstract. The article presents the laboratory and industrial research and the analysis of the impact of modification of refractory concretes and products with finely-dispersed nanomaterials. The dependence of the structural and physical-mechanical characteristics of mullite-corundum refractories of different types (for ladle furnace lances and for supporting blocks of rolling mill carriages) on nanomodification by electromagnetic homogenization is shown. The possibility of using a vortex electromagnetic homogenizer to significantly improve the performance of refractories is shown. In addition, the efficiency of using nanodispersed silicon dioxide is established and its optimal concentration is determined.

1. Introduction

Searching for the possibilities to reduce material costs is one of the key focus areas for steelmaking companies. In particular, improving performance of refractories is an important task both from the perspective of practical application and of metal science. Solving this task will make it possible not only to increase the quality of manufactured products, but also to improve such significant parameters as productivity and profitability. In order to solve the problem, energy-efficient modification of the available refractory materials with the use of various finely-dispersed modifying additives may be applied. At present, such approach is used in many industries. The main remaining problems are selection of the type of a modifying additive and its quantity, and achievement of a high degree of binding material homogeneity.

It is known that small additions (below 1 weight percent) of nanoparticles allow achieving a significant improvement of performance characteristics of products due to their influence on the structure formation processes. It should be noted that the existing methods of modifying high-alumina binding materials, such as mixed grinding in ball mills and nanoparticles injection with the use of suspensions do not give the desired outcome for various reasons: long processing time, particle-particle aggregation, segmentation and subsequent lamination, which is proved by a number of recent works in this area all over the world. In the recent researches not only alumina and silica nanoparticles, but also such complex compounds as zirconium silicate $ZrSiO_4$ [1–5] have been used. With this in mind, a method of vortex electromagnetic homogenization of the ground stock with the use of a finely-dispersed silicon dioxide additive is applied in this work.

Samples obtained during the work were examined with the use of up-to-date techniques, including: measurement of specific surface area by low-temperature nitrogen adsorption, nondestructive and destructive testing methods to measure ultimate compression strength, density determination and thermocycling.

All investigations were carried out with the use of the vortex electromagnetic homogenizer developed and manufactured by the Research and Production Enterprise Intor (OOO NPP Intor) in collaboration with the department of functional nanosystems and high-temperature materials of the National University of Science and Technology MISIS. This paper presents calculations of operational parameters for VEH-80 (Vortex Electromagnetic Homogenizer-80). Optimal process parameters for operation of the electromagnetic homogenizer, particularly quantity and maximum permissible weight of ferromagnetic rods, the loading density in the active area of the homogenizer, were determined. Optimal processing conditions and their influence on the structure of ceramic samples were established.

2. Input materials

High-alumina calcium aluminate-based binder is used as an input material. This binder is used for manufacture of refractories with addition of other components. The test sample of high-alumina cement is composed of aluminium oxide (69.5 %), calcium monoaluminate (23.5 %) and powder mullite (7.0 %). Particle-size distribution is shown in Figure 1.

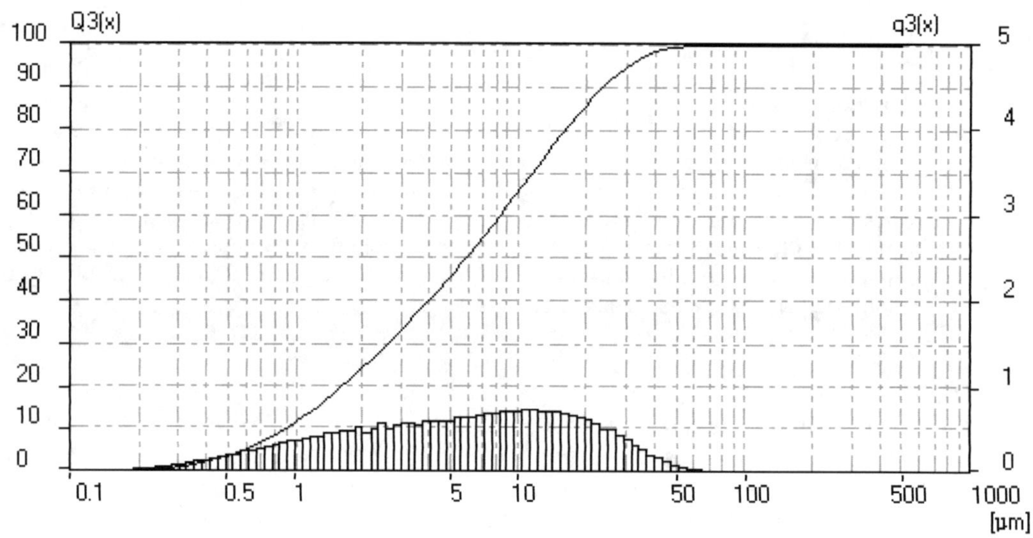

Figure 1. Particle-size distribution of the binder test sample (AIM grade cement).

Average size of the binder particles is within the range from 5 to 10 micrometers. The binder composition based on the results of the X-ray phase analysis is given in Table 1. The content of components may vary within 3-4 weight percent depending on the batch.

Commercially available hydrophilic AEROSIL® 300 with an average particle size of 10 nm is used as nano-sized SiO_2 in this research. This material is industrially produced by hydrolysis of a volatile silicon compound in oxy-hydrogen flame. Fine-grain structure of SiO_2 is formed during this process. Particles of AEROSIL® 300 may vary from 7 to 40 nm.

IOP Conf. Series: Materials Science and Engineering **718** (2020) 012018 doi:10.1088/1757-899X/718/1/012018

Table 1. Chemical composition of the binder.

Main components	Content, weight fraction, %
Al_2O_3	37.5 – 41
CaO	35.5 – 39
SiO_2	3.5 – 5.5
Fe_2O_3	13.0 – 17.5
MgO	Below 1.5
TiO_2	Below 4

Special chemical properties of this material are due to the presence of siloxane and silanol groups on the surface of particles. Siloxane groups predominate, which determines inertness of synthetic silicic acid, while silanol groups give the material hydrophilic properties.

Examination of this material with the use of a scanning electron microscope showed that is was composed of aggregates of different sizes ranging from 0.5 to 40 µm. Silicon dioxide structure is shown in Figure 2.

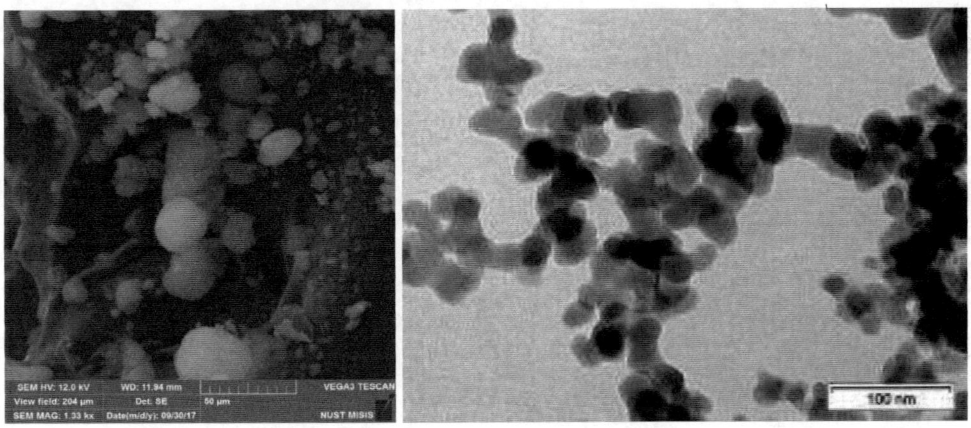

Figure 2. AEROSIL® 300 micrograph.

The aggregates shown in the micrographs are, in their turn, composed of individual particles from 7 to 40 µm in size, which is proved by transmission electron microscopy. The specific surface area of this material was measured by the low-temperature nitrogen adsorption method. The measured value was 258 m^2/g.

3. Nanomodification device

The vortex electromagnetic homogenizer (VEH) is designed to make homogeneous mixtures of ceramic and other nonmagnetic materials with the use of magnetic agitators and rotating electromagnetic field.

The homogenizer VEH ensures high-performance mixing and mechanical activation of ceramic materials due to fast rotation in the running magnetic field of ferromagnetic rods, which ensure mixing in fluidized bed.

Figure 3 shows pictures of the homogenizers VEH (VEH-160 – semi-industrial unit, VEH-80 – laboratory unit)

IOP Conf. Series: Materials Science and Engineering **718** (2020) 012018 doi:10.1088/1757-899X/718/1/012018

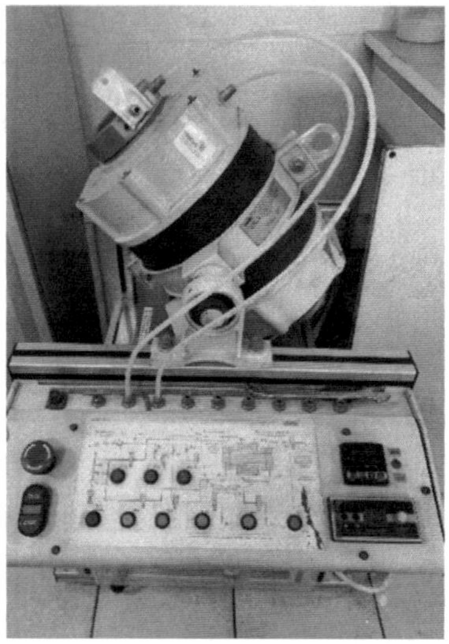

Figure 3. Pictures of the semi-industrial unit VEH-160 (on the left) and the laboratory unit VEH-80 (on the right).

To process the binder with the use of a finely-dispersed modifying additive, conditions that had been previously determined as optimal and allowing achievement of the most homogeneous particle-size distribution in the volume of material were selected.

A laboratory set of samples with varying content of Aerosil ® was produced with the use of this method. Based on the obtained results, a commercial batch of refractory concrete was produced. The content of the nano-sized modifying additive varied within the range from 0.4 to 1.2 weight percent of the binder weight.

4. Results and discussion

Based on the results of the research, optimal parameters of the modification of refractories for different applications were determined in terms of their influence on physical and chemical properties of refractories. Table 2 shows results of examination of physical and mechanical properties of refractory concretes depending on the nanomodifier amount.

Table 2. Apparent density and tensile strength of samples of concrete for supporting blocks of the rolling mill carriages.

Sample	Apparent density after drying, g/cm^3	Apparent density after burning, g/cm^3	Tensile strength after drying, MPa	Tensile strength after burning, MPa
AIM	2.65	2.59	96.9	138.5
AIM VEG	2.57	2.50	82.1	86.9
0.4% SiO_2	2.63	2.59	92.1	107.8
0.5% SiO_2	2.66	2.59	99.5	147.6
0.6% SiO_2	2.63	2.60	95.9	122.7
0.7% SiO_2	2.66	2.61	88.3	106.4
0.8% SiO_2	**2.65**	**2.62**	**115.7**	**171.4**
0.9% SiO_2	2.68	2.64	100.9	173.9
1% SiO_2	2.63	2.60	118.5	162.1

Optimal concentration of SiO_2 nanoparticles in the samples of concrete amounted to 0.8 weight percent with reference to the weight of binding material (0.04 weight percent of the weight of concrete). A proved increase of the true pycnometric specific gravity of the laboratory samples of 11 and 20 percent for concretes for blocks and lances respectively was achieved. Strength of laboratory samples after burning increased by 23.8% (for concrete for blocks) and 32.1% (for concrete for lances). Based on the obtained data, optimal process conditions of modification with the use of the vortex electromagnetic homogenizer were proposed for manufacture of trail batches of concretes for supporting blocks of the continuous reheating furnace carriages at the Plate Mill (Rolling Mill-3) (North Cast ABL 70.022) and for manufacture of unshaped concretes for lances for top argon blowing of metal in the ladle of the ladle-furnace North Cast BL 80.023/01. Manufacture of the trial batch and analysis of the results were part of the research conducted in collaboration with PAO Severstal.

The results of the performance tests of the supporting blocks for carriages are given in Table 3.

Table 3. Results of the performance tests of the supporting blocks for carriages.

Type of supporting block	Average time in operation of supporting blocks, days	Average time in operation of reference blocks for comparison, days
412	106	64
414	83	63
415	136	72
416	86	78

The average life calculated by averaging the values of all 12 nanomodified blocks amounted to 102 days (a 41 % increase).

The results of the performance tests of the ladle-furnace lances are given in Table 4.

Table 4. Results of the performance tests of the ladle-furnace lances.

Number of test lance	Life, minutes
1	207
2	183
3	117
Average value	169

The average life of the lances amounted to 133 minutes, which means a 27 % increase compared to the average value.

5. Conclusions

Optimal parameters of the modification of refractories for different applications were determined in terms of their influence on physical and chemical properties of refractories. Optimal concentration of SiO_2 nanoparticles in the samples of concrete amounted to 0.8 weight percent with reference to the weight of binding material both for concrete for lances and concrete for carriage blocks.

An increase of the true pycnometric specific gravity of 11 and 20 percent for concretes for blocks and lances respectively was achieved. Strength of final products after burning increased by 23.8 % (for concrete for blocks) and 32.1% (for concrete for lances).

The results of the conducted research and tests suggest that it is possible to improve the performance parameters of refractory products made of unshaped mullite-corundum mixtures by at least 20% due to nanomidification. Industrial application of this technology will ensure a significant economic benefit due to reduction of costs related to purchasing refractory materials and increase of interrepair cycles.

References

[1] Shahraki A, Ghasemi-Kahrizsangi S and Nemati A.2017 *Materials Chemistry and Physics* **198** 354–359

[2] Shiri S, Abbasi M H, Monshi A and Karimzadeh F 2014 *Composites Part B: Engineering* **56** 30–33

[3] Ghasemi-Kahrizsangi S, Karamian E and Dehsheikh H G 2017 *Ceramics International* **43** 13932–13937

[4] Singh A K and Sarkar R 2016 *Ceramics International* **42** 12937–12945

[5] Yanturina R A, Trofimov B Ya and Ahmedjanov R M 2017 *Procedia Engineering* **206** 869–874

IOP Conf. Series: Materials Science and Engineering **718** (2020) 012019 doi:10.1088/1757-899X/718/1/012019

Effective Materials from Recycled Aluminum Waste for Steel Ladle Treatment

L V Tribushevskiy[1], **G A Rumyantseva**[1] **and B M Nemenenok**[1]

[1] Belarusian National Technical University, 65 Nezavisimosty Avenue, Minsk, 220013 Belarus

E-mail: iakozhevnikova@chsu.ru

Abstract. The paper presents the results of research on the use of recycled aluminum waste (RAW) for steel ladle treatment. It has been established that simultaneously with the deoxidation of refining slags before desulfurization of steel, their liquid mobility increases, which eliminates the need for using fluorspar for these purposes. A thermodynamic analysis of possible reactions of formation of aluminates and calcium aluminosilicates in the refining slag has been carried out. The composition of briquettes based on RAW for liquefaction of refining slags is proposed.

To obtain high-quality steel at metallurgical plants, steel ladle treatment is carried out, which provides the required chemical composition, melt temperature, and a decrease in the number of non-metallic inclusions. An analysis of the dynamics of changes in requirements for the content of impurities in steels shows that from 1970 to 2010 the permissible sulfur concentration decreased significantly (from 0.03% to 0.003%), and for A-class steel, even more strict limits were set (0.002%). According to forecasts, by 2020, the steel pollution of this class with sulfur should not exceed 0.0015% [1]. Since 2014, for the pipes and fittings, the normative and technical documentation of Gazprom reflects the acceptable pollution standards for sulfide, oxide and silicate inclusions according to the average and maximum scores [2].

Researchers note the great potential for processing steel in the ladle using active refining slags. However, the thermodynamic and kinetic abilities of such a refining are used only by 10-12%. The efficiency of steel refining is influenced by the chemical composition of the slag, its sulfide capacity, slag oxidation, temperature treatment conditions, slag viscosity and process hydrodynamics [3].

The refining ability of slags mostly depends on their liquid mobility, since desulphurization and deoxidation processes develop at the slag-melt boundry. Most enterprises use fluorite (fluorspar) CaF2 in order to liquefy slag [4], but it is an environmentally hazardous substance, has a high cost and damages the lining of a steel-teeming ladle.

When conducting experimental melts at the Belarus Steel Works using aluminum deoxidizing agents obtained on the basis of recycled aluminum waste, it was found that aluminum deoxidizing agents positively affect the degree of steel desulphurization and the liquid mobility of the refining slag of the ladle furnace.

IOP Conf. Series: Materials Science and Engineering **718** (2020) 012019 doi:10.1088/1757-899X/718/1/012019

The addition of aluminum deoxidizing agents in an amount of 100 kg (or so) ensured an increase in slag mobility, as in the case of slag processing using both CaF_2 and deoxidizing agents.

The authors of [4] also note the positive effect of A12O3 on slag liquefaction and increase the durability of the lining of the steel-teeming ladle. They also recommend to use Rantal 50GR alumina-containing material (produced by RANTAL, St. Petersburg), alumina-containing briquettes (produced by ROIS, Yekaterinburg) and slag-forming material MShA Frado-1.

Thus, introducing additives containing CaO and Al_2O_3 into the refining slags, one can control the basicity and liquid mobility of the slag, as well as eliminate the need to use fluorspar (CaF_2) [4-6]. It has been established that the content of 20–25% of Al_2O_3 in slags eliminates the need for fluorspar to liquefy them. It should also be considered that aluminates with a high content of calcium (C_3A, C_5A_3) have a higher ability to absorb CaS than aluminates with a high content of alumina (CA, CA_2). In this case, the solubility of CaS significantly depends on temperature [7].

To assess the potential possibility of reactions between Al_2O_3 and the components of refining slag to form calcium aluminates and aluminosilicates, the Gibbs energy delta (change) was calculated for 16 reactions in the temperature range of 400–1873 K [5, 6]

1) $(4/3)\ 3CaO{\cdot}Al_2O_3 + Al_2O_3 = (1/3)12CaO{\cdot}7Al_2O_3$;
2) $12/7CaO + Al_2O_3 = (1/7)12CaO{\cdot}7Al_2O_3$;
3) $3CaO + Al_2O_3 = 3CaO{\cdot}Al_2O_3$;
4) $CaO + Al_2O_3 = CaO{\cdot}Al_2O_3$;
5) $(1/2)3CaO{\cdot}Al_2O_3 + Al_2O_3 = (3/2)CaO{\cdot}Al_2O_3$;
6) $1/2CaO + Al_2O_3 = (1/2)\ CaO{\cdot}2Al_2O_3$;
7) $(1/5)3CaO{\cdot}Al_2O_3 + Al_2O_3 = (3/5)\ CaO{\cdot}2Al_2O_3$;
8) $CaO{\cdot}Al_2O_3 + Al_2O_3 = CaO{\cdot}2Al_2O_3$;
9) $(1/17)12CaO{\cdot}7Al_2O_3 + Al_2O_3 = (12/17)\ CaO{\cdot}2Al_2O_3$;
10) $(1/5)12CaO{\cdot}7Al_2O_3 + Al_2O_3 = (12/5)\ CaO{\cdot}Al_2O_3$;
11) $Al_2O_3 + 2CaO + SiO_2 = 2CaO{\cdot}Al_2O_3{\cdot}SiO_2$;
12) $Al_2O_3 + CaO + SiO_2 = CaO{\cdot}Al_2O_3{\cdot}SiO_2$;
13) $1/3Al_2O_3 + CaO + SiO_2 = (1/3)3CaO{\cdot}Al_2O_3{\cdot}3SiO_2$;
14) $Al_2O_3 + 2CaO + SiO_2 = 2CaO{\cdot}Al_2O_3{\cdot}SiO_2$;
15) $1/2Al_2O_3 + 1/2CaO + SiO_2 = (1/2)\ CaO{\cdot}Al_2O_3{\cdot}2SiO_2$;
16) $CaO{\cdot}SiO_2 + CaO{\cdot}Al_2O_3 = 2CaO{\cdot}Al_2O_3{\cdot}SiO_2$.

Figures 1 and 2 show the Gibbs energy delta (change) of these reactions at a temperature of 1873 K.

From the presented calculations it is clear that it is theoretically possible for all 16 reactions to occur due to negative ΔGt values. However, it should be noted that as a result of reactions' behavior, fairly high-melting compounds are formed: $3CaO{\cdot}Al_2O_3$ (t_{melt}=1535°C); $CaO{\cdot}Al_2O_3$(t_{melt}=1610 °C); $CaO{\cdot}2Al_2O_3$ (t_{melt}=1765 °C); $CaO{\cdot}Al_2O_3{\cdot}2SiO_2$ (t_{melt}=1553° C); $2CaO{\cdot}Al_2O_3{\cdot}SiO_2$ (t_{melt}=1593 °C); $12CaO{\cdot}7Al_2O_3$ (t_{melt} =1527 °C) [5, 8]. Considering that refining slags are multicomponent, it is also possible to form the $4MgO{\cdot}5Al_2O_3{\cdot}2SiO_2$ (t_{melt} = 1475 °C) compound. Therefore, the effect of increasing the liquid mobility of refining slags with the addition of Al_2O_3 should obviously be associated with the possibility of the formation of low-melting eutectics.

Thus, eutectics with melting points of 1170, 1265, 1310, 1345, 1335 °C are formed in the CaO – Al_2O_3 – SiO_2 system [5, 8].

For the MgO – CaO – Al_2O_3 – SiO_2 system, three eutectics with low melting points can be indicated [8]:
5 % MgO–48,8 % CaO–41,5 % Al_2O_3–5 % SiO_2 (t_{melt}=1295 °C);
2,5 % MgO–23,8 % CaO–14,2 %Al_2O_3–59 % SiO_2 (t_{melt}=1160 °C);
9,4 % MgO–10,2 % CaO–18,5 %Al_2O_3–61,9 % SiO_2 (t_{melt}=1222 °C).

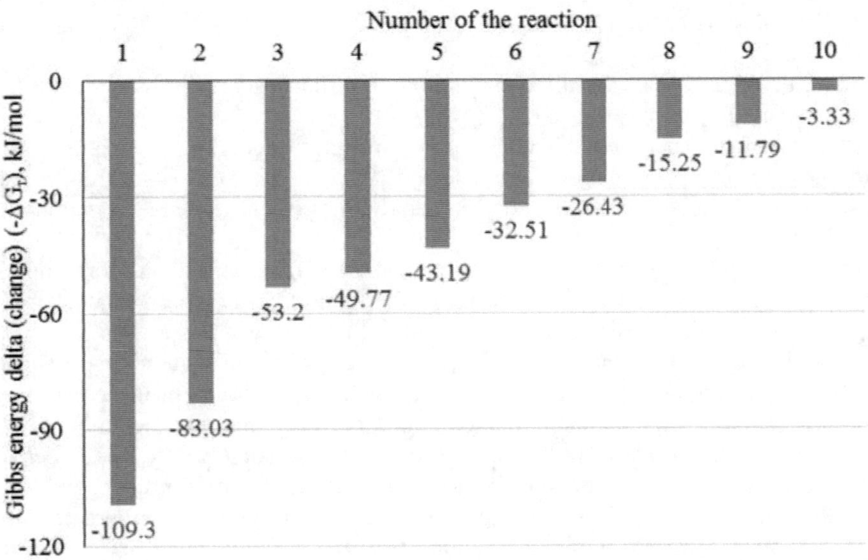

Figure 1. Gibbs energy delta (change) of reactions of calcium aluminates formation at a temperature of 1873 K (reactions 1-10).

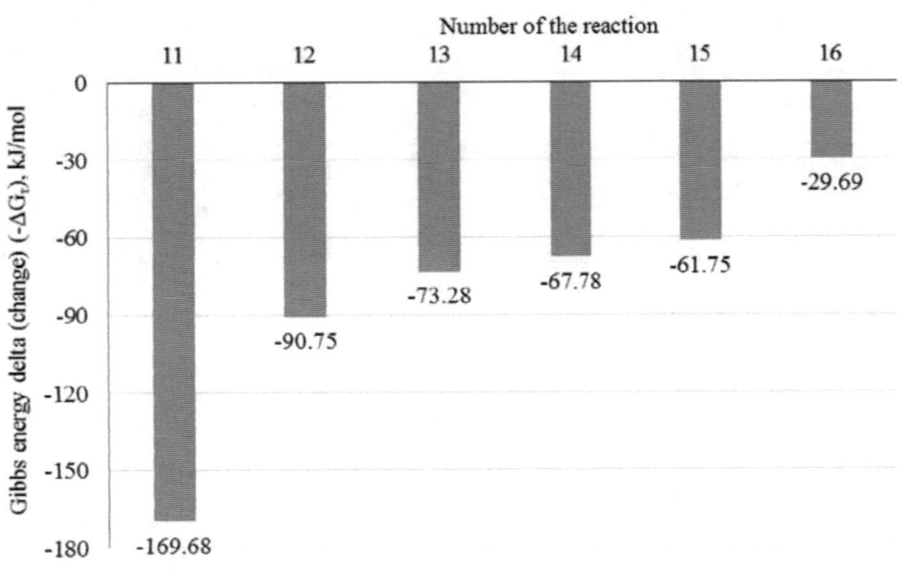

Figure 2. Gibbs energy delta (change) of reactions of calcium aluminosilicates formation at a temperature of 1873 K (reactions 11-16).

IOP Conf. Series: Materials Science and Engineering **718** (2020) 012019 doi:10.1088/1757-899X/718/1/012019

Eutectics with melting points of 1355 and 1365 °C are also formed in the MgO – Al_2O_3 – SiO_2 system [8].

The adjustment of Al_2O_3 content in the composition of refining slags in order to liquefy them can be carried out using recycled aluminum waste [5].

For the production of an experimental batch of a refining slags thinner, recycled aluminum waste stored at an open storage area of NPF Metallon were selected [9]. In order to separate the prills of aluminum, aluminum wastes were sieved through a sieve with square holes (with a side of 5.6 mm and a diagonal of the hole of about 8 mm).

Samples were taken from the obtained screenings of the different horizons for a detailed chemical analysis, which was carried out at the testing center of the Institute of Powder Metallurgy of the National Academy of Sciences in Belarus. Screening of recycled aluminum waste contains alumina (as the main component), a small amount of magnesium oxides, silicon and iron. Subsequent analysis showed that the content of the 0-3 mm fraction in the screening is 78-80%. This fraction is of the greatest interest for the production of high-alumina fluxes and the liquefaction of ladle slags in steel production.

To separate the aluminum prills from the screened slag, it was subjected to vortex separation using a conveyor belt with a multi-pole magnetic rotor. During the rotation of a multi-pole magnetic system, vortex currents were induced in metal particles, which, in turn, created a magnetic field opposite in direction to the rotor magnetic system. As a result of the interaction of magnetic fields, metal particles contained in the screened slag were removed from the stream and separated from the non-conductive fraction [9].

For production tests, an experimental batch of a thinner in the form of briquettes based on recycled aluminum waste with the addition of 40% CaO and a binding substance was made. Briquettes were obtained using a roller press. Data on the chemical composition and distribution coefficient of sulfur of refining slag before and after the addition of a thinner are shown in table 1.

Table 1. Chemical composition of refining slag and sulfur distribution coefficient (L_s).

Slag characteristics	Slag's chemical composition (mass fraction of oxides), %					Sulfur distribution coefficient (L_s)
	CaO	SiO_2	MgO	Al_2O_3	FeO	
Refining (initial)	46.5	24.4	9.8	12.6	2.6	123.4
Refining (after the thinner is added)	52.7	19.3	8.7	18.4	0.9	146.2

From table 1 it is seen that the thinning additives lead to an increase in the content of CaO and Al_2O_3 in the slag, with a noticeable decrease in iron oxides, apparently due to the content of metal aluminum prills in the additive. It should also be noted that the sulfur distribution coefficient increased from 123.4 to 146.2, which indicates the best desulfurizing ability of slag.

The addition of briquettes to the surface of the refining slag of a 100-ton steel-teeming ladle in an amount of 250 kg also provided a noticeable liquefaction of slag, which is confirmed by an increase in the mass of slag in the immersion viscometer by 70% compared to the refining slag of the initial state.

The variety of waste aluminum alloys to be processed requires careful preliminary preparation and sorting in order to obtain products with high added value. The most common wastes from aluminum and its alloys contain oiled paper, heat-insulating foam, plastic, ceramics and glass. With the introduction of compulsory sorting of household waste, the share of scrap and aluminum waste of grade 4 (class G) including bottle caps, vodka dispensers, metal cans for drinks, for the production of which food-grade aluminum alloys are used, has increased.

For the processing of such wastes, a scheme for their preparation and melting was proposed, with the production of aluminum granules using the "dry" technology.

At the first stage of preparation, recycled aluminum wastes were processed in a hammermill for glass breaking. Subsequently, the polymer-metal concentrate underwent magnetic separation in the unit with permanent magnets and the sorted magnetic component was used as part of the furnace-charge in the production of pig deoxidizing agent. The non-magnetic part of the waste was sent to a shredder for grinding to a fraction of 10-15 mm, which was subsequently subjected to magnetic and vortex separation. Using the vortex separation, a non-magnetic metal fraction was extracted, which is further used as a component of the furnace-charge during melting in a short-flame rotary furnace. The non-metallic fraction, consisting predominantly of plastic, was subject to processing to produce granules of recycled plastic sold to customers.

After the crushed and refined aluminum was melted in the short-flame rotary furnace, the melt was poured into a KROWNMATIC dosing furnace, where it was kept at a temperature of 740-760 °C for three hours, and then the metal was fed through a heated tray to the CENTAUR unit to obtain aluminum granules using "dry" technology.

Using such schemes for the preparation of recycled aluminum waste, an experimental batch of aluminum granules AB87 (weighting 5 tons) was manufactured using dry technology and delivered to the Moldova Steel Works for production tests. The granules were predominantly spherical in shape and loaded into the silo of the ASIS injection unit.

Aluminum granules were injected into a stream of metal at the time of release from EAF in accordance with the requirements of current technology. Experimental granules were used instead of crushed aluminum in the production of a low-carbon steel with a consumption coefficient of 0.96 relative to the norm for the material used.

The obtained technological indicators of steel ladle treatment for check and experimental melts had close values both in the specific consumption of pure aluminum delivered to the steel ladle with deoxidizers and in the content of carbon, silicon, aluminum, and sulfur in continuously cast billets.

Conclusions

Thus, non-waste technology for the processing of oxidized aluminum waste makes it possible to obtain effective materials for steel ladle treatment in the form of aluminum ingots, pyramids, granules, deoxidizing mixtures and refiners for refining slag. By doing so, it is possible to provide the necessary liquid mobility of refining slag and increase its desulphurizing ability, reduce the impact on the environment, eliminating the use of CaF_2 and the disposal of waste slag and, as a result, obtain significant environmental economic and environmental effect.

References

[1] Elansky D G 2016 *Steel* **11** 21–29
[2] Safronov A A, Golovin V V, Belokozovich Yu B, Matus V M, Ioffe A V and Movchan M A 2016 Steel **6** 22–27
[3] Melnik S G 2017 *Proc. of XIII International Scientific and Practical Conf. Casting. Metallurgy* (Ukraine: Zaporozhye) pp 354–356
[4] Metelkin A A, Sheshukov O Yu, Nekrasov I V and Shevchenko O I 2015 *Improving the durability of the lining of steel ladle treatment units* (Nizhny Tagil: Nizhny Tagil Technological Institute of Nizhny Tagil (the branch of Ural Federal University)) p 144
[5] Tribushevskiy L V, Nemenenok B M, Rumyantseva G A and Kulik M A 2018 Casting and metallurgy **1** 100–105
[6] Tribushevskiy L V, Nemenenok B M and Rumyantseva G A 2018 *Proc. of XXVI International Scientific and Practical Conf. Casthouse production and metallurgy* (Belarus: Minsk) pp 115–117

IOP Conf. Series: Materials Science and Engineering **718** (2020) 012019 doi:10.1088/1757-899X/718/1/012019

[7] Dyudkin D A, Kiselenko V V and Bat S B 2007 *Steel* **8** 25–29
[8] Bobkova N M 2007 *Physical chemistry of higher-melting-point non-metallic and silicate materials* (Minsk: Vyshejshaya shkola) p 301
[9] Tribushevskiy L V, Nemenenok B M and Rumyantseva G A 2019 *Proc. of XV International Scientific and Practical Conf. Casting. Metallurgy* (Ukraine: Zaporozhye) pp 205–206

IOP Conf. Series: Materials Science and Engineering **718** (2020) 012020 doi:10.1088/1757-899X/718/1/012020

2020 Technology and Operation of a Hot Rolling Mill

D Vanderschueren[1]

[1] Steelmasters bvba, Berchemstraat 14, Roosdaal, 1760, Belgium

E-mail: dirk.vanderschueren@steelmasters.be

Abstract. The state-of-the-art of both technology and operational practice of a hot rolling mill are discussed. The focus is first on the control of entry temperature in the finish mill. Transfer bar cooling and induction heating are presented. Next, the safe keeping in the finishing mill of surface quality, crown and flatness are discussed. Important tools to achieve these properties are lubrication, friction measurement, roll cooling, inter stand and skin cooling, bending, shifting and prediction of roll forces. The third and final topic is the run- out table cooling. Ultra-fast cooling is compared with laminar cooling. The cooling strategies and their importance to realise the required micro-structure and properties are discussed for hot rolled Dual Phase steel, pipe line grades and rebar.

The first major operation in the hot rolling mill is the reheating of slab or billet. There is a lot to be said about reheating furnaces, both about their technology and their exploitation. The reheating operation is very important for the environment, cost and metal yield. About 30% of the variable cost to transfer a slab into a hot rolled coil is spent on natural gas for the reheating furnaces. That is why the attention that is paid to the reheating furnace is very much justified. The energy requirement is minimised by loading the slabs at high temperature (= hot charging) and keeping the necessary reheating temperature as low as possible. The latter is done by reducing the heat loss in the line. Radiation losses are reduced using heat panels. When applying water spraying, as in the case of descaling, attention should be paid to convection losses too.

Although reheating furnaces are the preferred way to bring the slabs to high temperature, they have a number of distinct drawbacks:

– Direct flame contact in the reheating furnace will oxidize the slab with typically 1% of material loss as a consequence.

– The inertia of a reheating furnace is high. Changing the temperature takes time. The discharge temperature can only be changed gradually. A reheating Furnace is not suitable for a precise, slab to slab adjustment of the discharging temperature in correspondence with the aimed entry and exit finishing mill temperature.

This lack of flexibility is a serious drawback. Because it is not possible to change the unloading temperature from slab to slab, consecutive slabs are heated to a temperature sufficiently high to accommodate the slab that needs the highest temperature discharge temperature. This is the slab that is going to be rolled to a small thickness (= high heat loss) and with a high finishing rolling temperature. As a consequence, the other slabs will be heated to a higher temperature than strictly necessary. Not only too much energy is put into them, the production pace is reduced because time must be given for the bar to cool down. Reheating energy is spent and rolling capacity is lost. That is a double loss.

IOP Conf. Series: Materials Science and Engineering **718** (2020) 012020 doi:10.1088/1757-899X/718/1/012020

Fortunately, tools exit to compensate this poor flexibility of the reheating furnace:
- Transfer bar cooling (Figure 1).
- Induction heating of the slab/billet or transfer bar.

Figure 1. Transfer bar cooling system installed at the hot strip mill of Tata Steel´s Port Talbot [2].

Whereas transfer bar cooling is installed on HSM's [1,2], induction heating is more frequently used for mini mills, that is, combined casting hot rolling lines [3, 4].

Transfer bar cooling is installed just after the roughening mill. To realise an average temperature drop of 50°C on a 32 mm roughing mill bar, the surface needs to be strongly cooled for several hundreds of degrees. Deforming a transfer bar with such inhomogeneous temperature profile will have a negative influence on the resulting microstructure. It is therefore necessary to allow time for temperature homogenisation before entering the finishing mill.

An alternative way to control the entry temperature of the finishing mill is by induction heating. The heat produced with induction can be produced through thickness in case of thick bars with longitudinal induction and at the surface on thin bars with transfer induction. Combined casting rolling lines are used to produce thin hot rolled grade, as thin as 0.6 mm [3]. Keeping the reduction capability in the finishing mill limited, this requires thin transfer bars. These thin transfer bars lose too much heat to finishing the rolling above AR3. To roll to such small thickness it is difficult to do without a transfer bar heating installing at the entry of the finishing mill.

In case of bloom or billet rolling, there are lines in which there is no more reheating furnace. It is replaced by an induction heater (Figure 2). The reported advantages are [4]:
1. a flexible solution for heating hot charged billets,
2. reduction of scale losses.

Figure 2. Induction heating for billets [4].

Next, the finishing mill is discussed. The tools that are used in the finishing mill to control the surface quality, the crown and flatness, thickness and width are:

1. lubrication, roll cooling, inter stand and skin cooling,
2. bending, shifting, prediction of roll forces and looper tension.

The first set of tools are there for the surface quality. Actually, their goal is to avoid finishing mill rolled in scale. In case the scale is not able to deform to the same extent as the steel, it is broken. This gives an incrustation of oxide particles. This principle is schematically presented in Figure 3 [5].

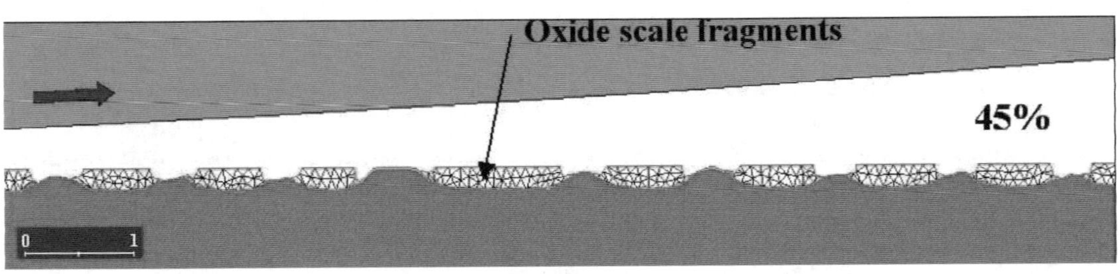

Figure 3. Scale breaking in the roll bite, simulation of 45% reduction.

Only in one, neutral point, the roll and the sheet have the same speed. In all other positions, there is relative, movement of roll compared to sheet. The lubrication reduces the friction between roll and sheet and, in this way, makes the deformation more homogeneous. This helps to avoid finishing mill rolled in scale. But, this is only one of the many influencing parameters. Other parameters are the size of the reduction, the temperature, the thickness and the nature of the scale and the condition of the work rolls. Traditionally, oil is sprayed on the work roll as a mixture of water and oil with the goal that

the oil sticks to the roll and remains there until the roll comes in contact with the sheet. It should be clear that this is not evident, and is influenced by many parameters. Recently, there have been a development to apply pure oil on the roll without water [7].
Several technologies are used to measure the result of the lubrication, that is the friction coëfficient between sheet and roll and how much it is reduced thanks to the lubrication. All of these measurements are indirect:

- slip forward
- torsion measurement
- comparison of predicted and measured rolling forces
- comparing rolling forces with and without lubrication

An example of the comparison of rolling force with and without lubrication is given in Figure 4.

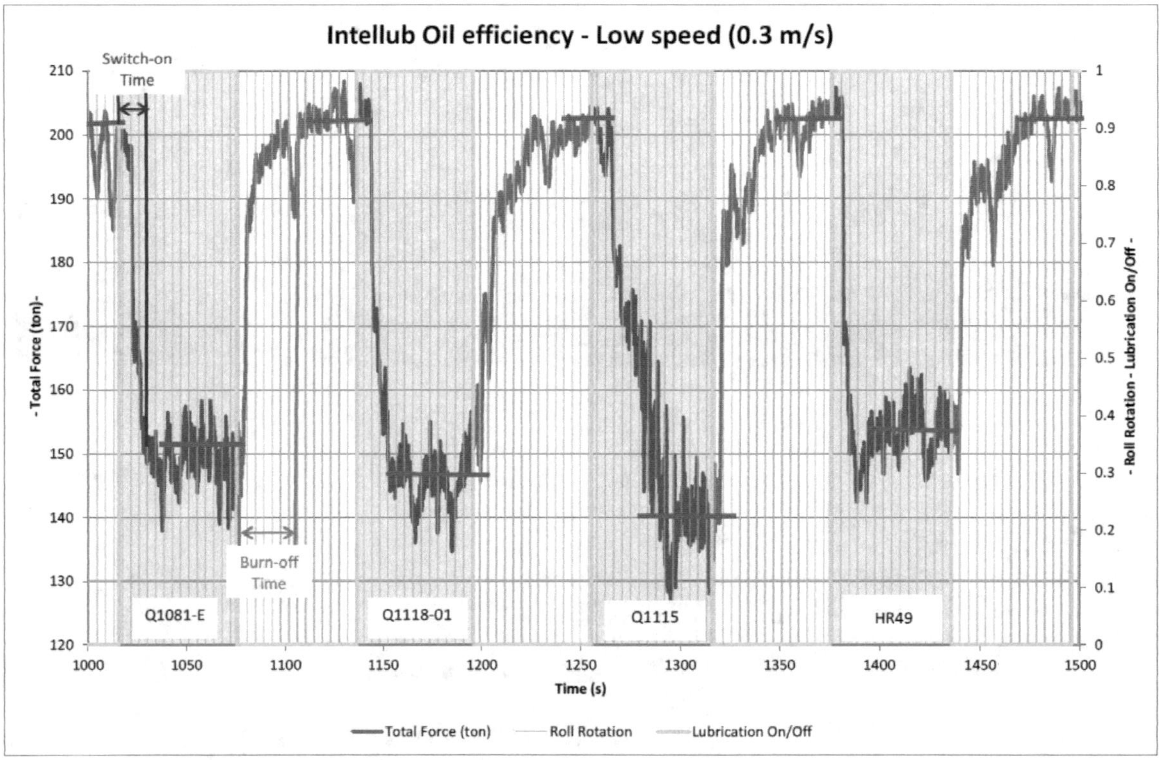

Figure 4. Rolling Forces with and without Lubrication [6].

Roll cooling is traditionally performed with sprays at high pressure. Performance improvement and reduction of cost are reported for a new type of low pressure cooling called High Turbulence Roll Cooling (HTRC) [8].This type of cooling module is presented in Figure 5.

The profile and flatness are determined by the shape of the roll gap. The differences in roll gaps from stand to stand must be sufficiently small so that the deformation can be realised without creating unflatness. The maximum allowable differences are determined experimentally and translated to a 'deformation trumpet'. An example is given in Figure 6.
The roll gap is adjusted using bending and shifting. However, everything start with the prediction of the roll forces. The roll gap set-up is chosen so that the expected opening and bending-out of the roll because of the rolling forces is compensated. The more precise the prediction of the rolling forces is, the better is the set-up and thus the thickness, crown and flatness at the head. The importance of the prediction of the rolling forces cannot be underestimated. It is important for the properties of the product and for the stability of the process (= avoidance of rolling incidents). A large part of the

planning rules are intended for this reason, to improve this prediction. This is done by limiting 'the changes' from one coil to the next. These differences are hardness, rolling temperature and dimensions. Out of lack of pure predictability, the help of adaption is required. The prediction of the rolling forces is all to more precise if the change is small.

Figure 5. High Turbulence Roll cooling module [8].

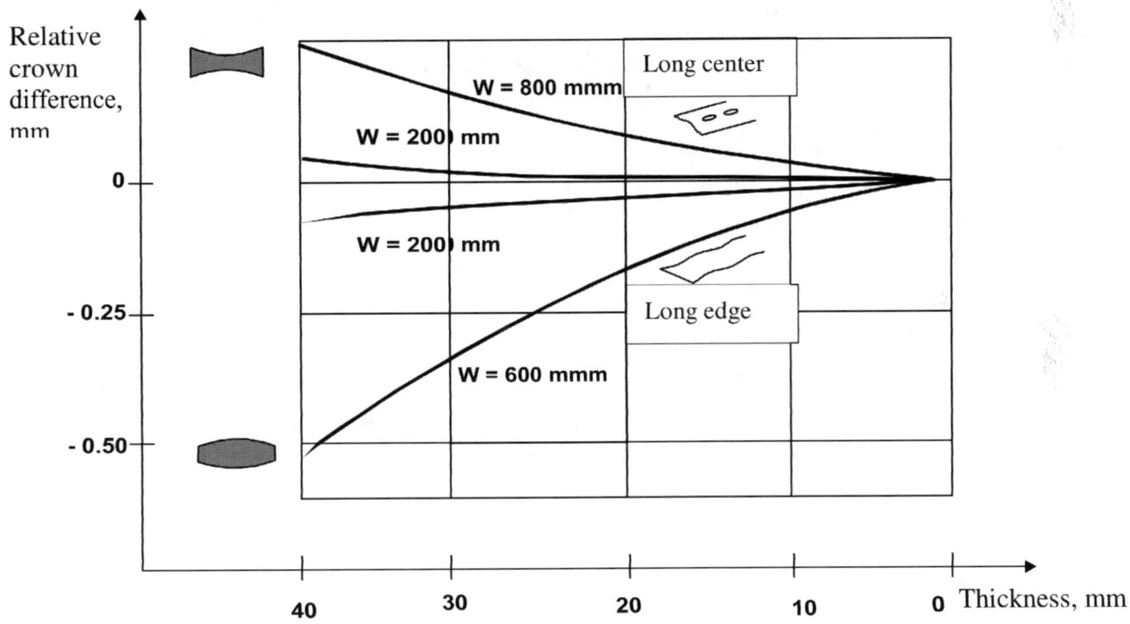

Figure 6. Allowable roll gap difference for two widths as function of thickness to avoid unflatness.

After the last deformation, the steel is cooled in 2 very different modes to room temperature. First there is a fast cooling on the run-out table. There is a typical temperature drop of 250°C in about 10s. Once the sheet is coiled, the cooling is slow. It takes 24h+ to realise the remaining 500°C to RT. The reason for the first fast cooling is the increase of the driving force for transformation and (cementite) precipitation. This is the route for the production of fine grained, non-ageing C-steel. It is argued that ultra-fast cooling has the advantage of creating more driving force and thus is able to reduce the amount of alloying elements. Please note that it is equally important to pay attention to the whole thermal history after the finishing mill. The distance, and thus the time between the last deformation and the start of accelerated cooling, is equally important.

The cooling on the run-out table is controlled precisely to influence the final microstructure. An example of this is the production of rebar with tempcore cooling [11,12] which is a strong, short cooling. As shown in Figure 7, this creates a heterogeneous cooling over the thickness. On the surface, there is a fast cooling, giving martensite. At the core, thecooling is slower, resulting in a ferrite-perlite structure. Moreover, the martensite is subsequently relieved by the heat of the core. As a result, a high strength steel with a YS > 500MPa and an uniform elongation of 10% can be obtained using a simple 0.2% C chemistry.

Also in sheet rolling, the cooling on the run-out-table is adapted according the requirements. An example is the production of hot rolled Dual Phase steel. A two-step cooling is used, that is:

- first water cooling is applied,
- second, an air cooling in which 85% ferrite is formed, is used and finally,
- again water cooling is used to promote the formation of martensite for the remaining 15%.

The goal of the first water cooling is to increase the driving force for ferrite transformation. The highest kinetics are at 650°C. In order to be able to stay as long as possible in this temperature region, the speed to go to it is as high as possible. However, the 'nose' of the transformation speed is flat. Nor the exact temperature, neither the high speed to go to it, are crucial.

The situation is different for the production of pipeline grades for both plate and sheet. The desired microstructure is bainite and the sheet/plate is thick. In this case, an accelerated cooling of the ultra-fast type has advantages [13] (Figure 8).

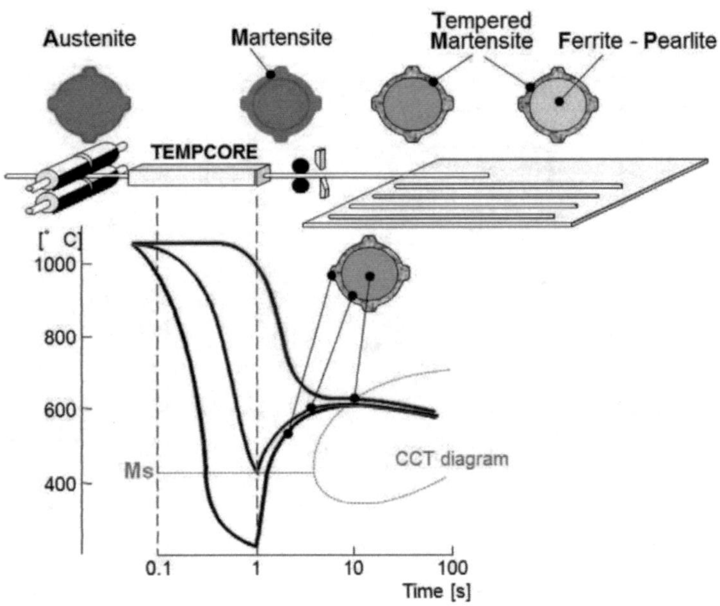

Figure 7. The principle of Tempcore (11).

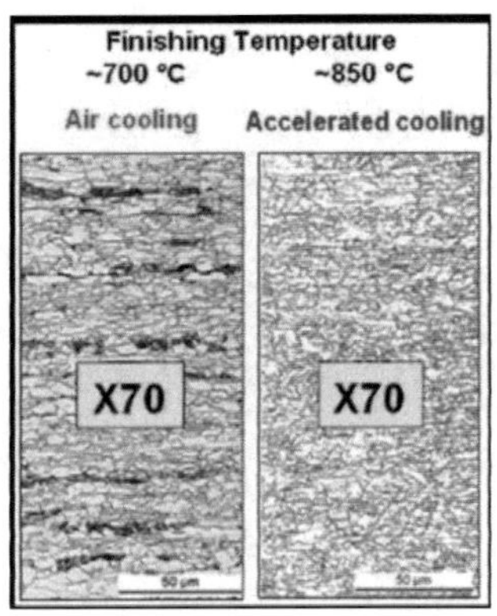

Figure 8. Grain structure under different cooling strategies [13].

Conclusion

Since the control of the entry temperature is important and the design of the line is not favourable for the thermomechanical needs of the products, tools have been developed to accommodate. There is bar cooling and induction heating.

Although pure oil application exists, Lubrication is generally applied as a mixture of water and oil. The efficiency of this oil is a concern and is influenced by many parameters. The prime goal of the lubrication is to reduce the rolled in finishing mill scale.

There are many tools to but a crucial point is the prediction of the rolling forces. The limited ability to do so has impact on set-up and is a hindrance for schedule free rolling.

Ultra-fast cooling and laminar cooling on the run-out table each have their strength and weaknesses and applications. The nature of the run-out table cooling for Dual Phase steel, Pipeline grades and rebar have been discussed.

References

[1] Bos B, Herman J C, Moerkerke J, Van Poecke P, De Zutter G, Declercq F De Paepe A and Simon P 2001 *ATS Paris Journées Sidérurgiques Internationales* (France: Paris)

[2] Pichler L, Seilinger A, Rimnac A, Lettner F, Weinzierl K, Milbredt J and White K 2019 *4th European Steel Technology and Application Days (ESTAD)* (Germany: Dusseldorf)

[3] Zahedi M, Lengauer T, Großeiber S, Linzer B, Jungbauer A, Umbrasko A and Thus A 2019 *4th European Steel Technology and Application Days (ESTAD)* (Germany: Dusseldorf)

[4] Arduni F and Gigante E 2019 *4th European Steel Technology and Application Days (ESTAD)* (Germany: Dusseldorf)

[5] Zhang M H; Liu B; Boréan J-L and Montmitonnet P 2011 *AIP Conf. Proc. 14th ESAFORM* (UK: Belfast) pp 333–38

[6] Malbrancke J and et al. *Final report of ECSC project* Increased mill capability and performances by the application of intelligent lubrication EUR 28973 2017

[7] Vervaet B, D Avedian D., Pesci C. 2011 *Steel Times International* **1**

[8] Uijtdeboreks H., Vervaet B, Lannoo G, Herman J-C, Vanderschueren D and Vanpoecke P 2010 *10th International Conference on steel rolling* (China: Beijing)

[9] Malbrancke J and et al. *Final report of ECSC project* Increased mill capability and performances by the application of intelligent lubrication EUR 28973 2017

[10] Vervaet B, D Avedian D and Pesci C 2010 *Revue de Métallurgie* **107** 237–244

[11] Economopoulos M, Respen Y, Lessel G. and Steffens G. 1975 *Metallurgical Reports CRM* Application of the Tempcore-Process to the Fabrication of High Yield Strength Concrete reinforcing **45** pp 3–19

[12] Noville J F 2015 *2nd European Steel Technology and Application Days (ESTAD)* (Germany: Dusseldorf)

[13] Krimpelstätter K, Pichler L, Rüdiger Döll R and Weinzierl K 2017 *Advanced Steel Plants for Production of Hot-Rolled Flat Products AIST* (USA: Warrendale) chapter 31 pp 399–412

IOP Conf. Series: Materials Science and Engineering **718** (2020) 012021 doi:10.1088/1757-899X/718/1/012021

Experimental Investigations of Physical Properties of Iron-Copper Alloy

K N Vdovin[1], G A Dubskiy[1], L G Egorova[1] and A A Nefedyev[1]

[1] Nosov Magnitogorsk State Technical University, 38 Lenin Street, Magnitogorsk, 455000, Russia

E-mail: vdovin@magtu.ru

Abstract. The paper describes the results of experiments aimed at obtaining iron-copper alloys with different iron content using electroslag remelting method. The obtained alloys of iron and copper were carefully examined to study their mechanical and physical properties. The article shows the results of the research work investigating the influence of iron on crystallization characteristics of the copper alloy. It was found, that iron-copper alloys can compete successfully with expensive copper-silver alloys, which are used to manufacture mold walls. The article offers the results of the experiments investigating hardness, wear resistance, thermal conductivity and temperature conductivity of the obtained iron-copper alloys; their casting characteristics were also studied.

1. General description of the problem under study

Modern industrial-scale process of continuous casting of steel puts forward new requirements to the materials used for mold walls of billet continuous casting machines; these materials must be able to withstand high temperatures for long periods of time, cyclic heat shocks and they must show minimum work wear during operation.

At present, there are a number of methods to solve this problem: changes in the design of the mold, application of coatings on the mold walls, copper alloying with various metals of the necessary content [1–4].

Copper alloys with a small amount of alloying elements are used to produce molds; introduction of such alloying elements improves the ultimate strength, yield strength, hardness and recrystallization temperature [3].

In order to provide high thermal conductivity (more than 90% of copper thermal conductivity), the concentration level of copper alloys must be very low. The increase in alloying element content does not result in any significant increase of the recrystallization temperature. Thus, the required combination of properties, such as high thermal conductivity and high temperature of softening in comparison with copper, can be achieved by alloying copper with a small amount of alloying elements [5].

Such alloys as Cu – Ag, Cu – Cr, Cu – Zr, Cu – Sn, Cu – Fe, Cu – Be, Cu – P, Cu – Fe – P, Cu – Sn – Zr, Cu – Cr – Zr and Cu – Co – Be are often used to manufacture mold walls.

To improve the tensile strength, thermal shock resistance and wear-resistance of a tube-type mold, it was suggested to introduce iron and phosphorus in certain proportions into the composition of copper alloy material of the mold. When 0.02-0.15 % of iron and 0.02-0.05 % of phosphorus is introduced into the alloy, a chemical compound of Fe_2P is formed, which strengthens the alloy.

As a result of comparative tests of a number of copper low alloys, it was recommended to use a nickel-copper-silver alloy as a material for mold walls, because this alloy showed higher mechanical properties, as compared with copper, copper-silver alloy and copper-zirconium alloy at high temperature as well as adequate thermal conductivity. Some properties of these alloys are given in Table 1.

Table 1. Properties of alloys used for mold manufacture.

Property	Copper		Copper-silver alloy		Copper-zirconium alloy		Nickel-copper-silver alloy	
	Test temperature, °C							
	20	300	20	300	20	300	20	300
Tensile strength, MPa	220	150	440	310	230	–	750	650
Yield strength, MPa	60	50	420	300	150	–	700	600
Hardness HB	500	380	1150	900	600	–	2400	2000
Thermal conductivity λ, W/(m·degree)	409	393	382	377	380	–	196	262

As it was mentioned above, alloying with P, Ag, Cr, Zr and Fe as well as cold deformation are used to improve mechanical properties of copper at high temperatures. Copper alloyed with 0.02-0.05 % of phosphorus is commonly used for mold tubes. Its recrystallization temperature is about 350 °C, and the main properties of this alloy are given in Table 2. Copper alloyed with chromium (about 0.65 %) and zirconium has the recrystallization temperature above 500 °C, however, in this case, thermal conductivity of copper decreases by 15-20 %.

Table 2. The main properties of Cu – P (0.02-0.05 %) alloy.

Property	Test temperature, °C					
	20	200	300	400	500	600
Tensile strength, MPa	360	310	270	120	90	60
Yield strength, MPa	350	300	260	50	30	20
Relative extension δ, %	15	15	15	45	62	65
Relative broadening ψ, %	70	75	75	92	99	100
Thermal expansion coefficient $\alpha \cdot 10^{-6}$,$°C^{-1}$	–	17.3	17.6	17.9	18.3	18.6

The main objective of this research work is to obtain an iron-copper alloy with improved performance characteristics, which will be used to manufacture mold walls for molds of billet continuous casting machines.

IOP Conf. Series: Materials Science and Engineering **718** (2020) 012021 doi:10.1088/1757-899X/718/1/012021

2. Results of experimental and theoretical investigations of obtaining iron-copper alloys

In the process of the alloy development, which will be used to produce mold walls, it should be taken into account that thermal conductivity of any material has a significant influence on the rate of outer skin growth of the solidifying ingot and on the operating temperature of mold walls.

In [3] it was suggested that the iron-copper alloy (containing 0.05-0.1 wt. % of Sn, 0.02-0.07 wt. % of Fe, 0.01-0.07 wt. % of P, the rest being Cu) may be used to produce molds. The mechanical properties of this alloy are 2-3 times higher than those of Copper 1 and Copper 2. It was found that at room temperature only 0.3 wt. % of iron can be solved in copper. At the same time, the obtained low copper alloy is an α-phase in the form of substitutional solid solution.

The cast specimens were used to study the microstructure of the alloys, the mechanical and thermophysical properties, which determine the qualitative characteristics of mold walls operating capacity.

The specimens containing iron within the solubility limit, that is, c = 0-0.3 wt. %, are characterized by a fine-grained structure. The majority of the grains have the form of a polygon with the average size about $d_m \approx 0.09$ mm. Besides, the average grain size decreases when iron content increases, which results in improvement of some mechanical properties of these alloys, for example, hardness. The hardness increases to 0.17 wt. % Fe.

Copper alloys with iron content $c \leq 0.3$ wt.% are single-phase crystalline substitutional solid solutions following Hume-Rothery's rule. Thus, the conducted metallographic investigations of low alloyed iron copper alloys confirm that the amount of the alloying element, that is iron, has a significant influence on the size and shape of grains in the alloy.

Hardness can change considerably with the changes in the structure, phase composition of the alloy, temperature and different kinds of thermal and mechanical treatment. The hardness of the examined alloys with iron content within the range of $0 < c \leq 0.3$ wt. % is higher than the hardness of the alloys containing over 0.3 wt.% of iron. This difference in hardness of the alloys is explained by the fact, that during crystallization the alloys obtained in the first concentration range form hard substitutional solid solutions in α–phase with finer grain, than in the second interval, where during crystallization of the alloys, (α + β)-phases are formed with coarser grain.

The results of the analysis of wear resistance are shown in Figure 1. Each point plotted on the graph is the average value obtained on the basis of five measurements.

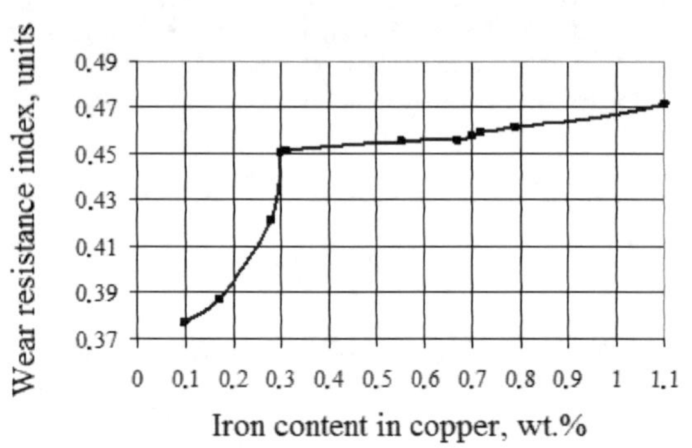

Figure 1. Relationship between the wear resistance index of low-alloyed Cu – Fe alloys and iron content.

From the diagram Figure 1, one can see that iron content increase to 0.3 wt. % results in sharp increase of wear resistance. If the iron content in copper increases even further, the value of wear resistance changes smoother. This can be explained by the fact that in the range of the concentration level up to 0.3

wt.%, saturation of solid solution of iron in copper reaches the solubility limit forming a substitutional solid solution, thus causing improvement of performance characteristics of the alloys, while further increase of the content does not seem reasonable.

The data of the experiments mentioned above show that alloying of copper with iron is reasonable to 0.3 wt. %. That is why the specimens № 1-6 were further examined in more detail. Differential scanning calorimetry method was used to determine the melting and crystallization temperatures. The results of the research are given in Table 3.

Table 3. Melting and crystallization temperatures of the examined alloys.

Specimen number	Fe content, wt. %	Melting, °C			Crystallization, °C		
		T_{np} (beginning)	T_{kp} (end)	$\Delta T = T_{kp}$-T_{np}	T_{np} (beginning)	T_{kk} (end)	$\Delta T = T_{kk} - T_{np}$
1	0.065	1053	1083	30	1068	1052	16
2	0.086	1045	1089	44	1063	1045	18
3	0.097	1044	1091	47	1060	1043	17
4	0.17	1066	1086	20	1052	1041	11
5	0.28	1072	1095	23	1039	1034	5
6	0.3	1074	1098	24	1037	1033	4

It follows from the analysis of the data from Table 3 showing the temperatures of the start and end of melting and crystallization for low alloyed copper alloys Cu – Fe that:

– the temperature range ΔT(the difference between liquidus and solidus temperatures during melting) has different values: at $0 < c \leq 0.17$ wt. % of Fe, it reaches its maximum ($\Delta T = 47$ °C, at $c = 0.097$ wt. %), and further it decreases to 20 °C at $c \geq 0.17$ wt. % and its value remains about the same;

– the temperature range of crystallization for the given iron content changes from the maximum of – 18 °C at $c = 0.086$ wt. % of Fe to 4 °C with the increase of the iron content in the alloy.

Before the experiments, the specimens with the preset iron content were annealed at the temperature of 650 °C for 8 hours. After that, they were divided into five sets. Each set of the specimens underwent deformation with different values of percent reduction, the step of reduction was 10%; they also underwent stepped annealing to the temperature of 400 °C. After annealing, the following parameters were measured:

1) the percentage change of resistance $\Delta R = R_0 - R_m$, where R_0 is metal resistance at $T = 300$ K; R_m – relative change in resistance after annealing;

2) Brinell hardness.

The results of the investigation of the relationship between $\Delta R/R_0$ and copper HB (copper 1) and the temperature for the preset value of the degree of plastic deformation are given in Figures 2 and 3.

From the analysis of resistance change in the process of annealing of the hardened Copper 1 (Figure 2), one can see that the value of $\Delta R/R_0$ in some temperature range does not change, then it drops monotonically from T_{np} (the temperature of the start of primary recrystallization) to T_{kp} (the temperature of the end of primary recrystallization), and after that when T increases, T_{kp} and $\Delta R/R_0$ reaches the value corresponding to the normalized (annealed) metal.

Figure 3 shows the dependence of Brinell hardness on the temperature of the copper specimens, which do not contain any iron and which underwent preliminary plastic deformation to a certain value ε.

The same experiments were carried out for low alloyed copper alloys with the following Fe, % wt. content: 0.097, 0.17, 0.28 and 0.3. The relationships between the hardness HB of these specimens and the reduction value are given in Figure 4. From the graphs in Figure 4, one can see that when the reduction value of the specimens with different iron content increases, their hardness also increases. Supposing that the dislocation concentration in the specimens with different iron content at the given reduction value is the same, then the increase in the hardness can be explained by a greater number of potential obstacles formed due to the impurity introduction. Their number increases with the increase in the concentration of the alloying element.

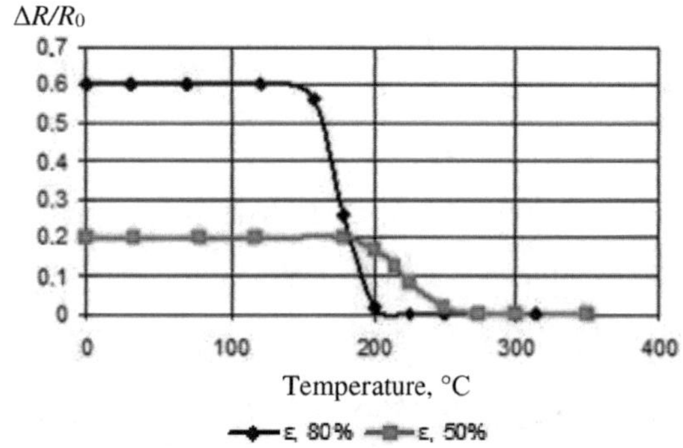

Figure 2. Dependence of relative electrical resistance of copper (Copper 1) on the temperature and the degree of plastic deformation ε.

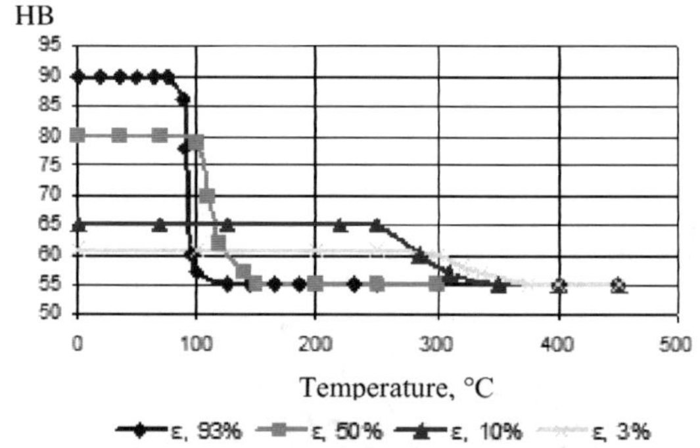

Figure 3. Hardness change of the strengthened Copper 1 in the process of annealing.

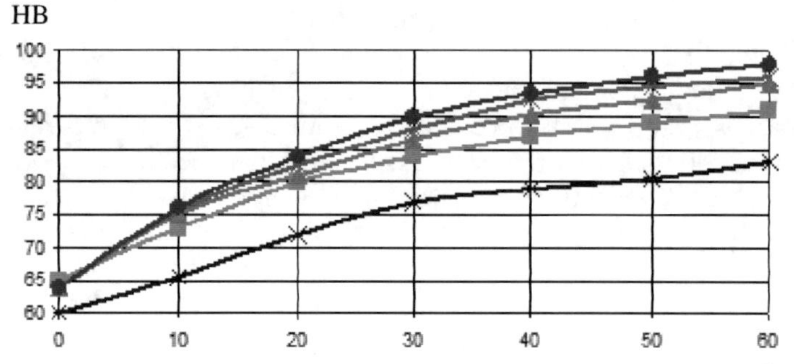

Figure 4. Relationship between the hardness of low alloyed copper alloys with different iron content and different degrees of plastic deformation ε.

Thus, as a result of examination of low alloys of Cu and Fe, and on the basis of the obtained results, the research group can offer a number of recommendations for practical implementation:

1. In accordance with the set of physical and mechanical properties obtained in the process of examination of Cu – Fe alloys, the material of choice in the process of mold walls production is the alloy with the iron content within the range of $0.17 \leq c \leq 0.25$ wt. % Fe.

2. In the process of mold walls production, they should undergo rolling of surface with a hard-alloy

IOP Conf. Series: Materials Science and Engineering **718** (2020) 012021 doi:10.1088/1757-899X/718/1/012021

roller providing mechanical stress, which is a bit higher than the yield strength of the used alloy. This will make it possible to improve the wear resistance of the wall by 10-15 % as compared with the unprocessed one.

The recrystallization temperature is not the only heat characteristic, which determines the durability of a mold. Its thermal and physical characteristics are just as important. That is why it is necessary to know how the iron content in copper influences its thermal conductivity and temperature conductivity.

To carry out the investigation of the thermophysical properties of low copper alloys, a number of specimens in the form of cylinders 12 mm in diameter and 16 mm in height were manufactured. Before placing them into the process chamber of the installation, all the specimens were annealed at t = 650 ^{0}C for 6 hours [9].

The results of the investigation of the dependence of the thermal conductivity and the temperature conductivity of copper and its low alloys on the temperature showed that thermal conductivity λ and temperature conductivity a of pure copper decrease monotonically when the temperature increases, while these characteristics of low alloys increase monotonically with the temperature increase.

In the examined low alloyed copper alloys, the alloying element (iron) changes significantly the energy-band structure, where apart from the 4s-band, 3d-band is formed, where copper is solvent, which can result in general increase in the thermal conductivity when the temperature increases.

Thus, the obtained results of the investigation of the thermal conductivity and the temperature conductivity of low alloy make it possible to come to the following conclusions:

1) At temperatures $T < 400$ ^{0}C, Cu – Fe alloys have the thermal conductivity and the temperature conductivity, which are 1.5-2 times lower, than the thermal conductivity and the temperature conductivity of pure copper.

2) When the temperature increases ($T > 600$ ^{0}C), these parameters become very similar to the parameters of pure copper and even exceed them slightly. The mold walls made of copper alloy containing 0.17-0.23 % of iron will have satisfactory thermal conductivity at the temperature of 250-300 ^{0}C.

Having studied the characteristics of hardness, wear resistance and thermal conductivity, the research group found that the best values of these parameters correspond to the alloy containing 0.16-0.18 wt.% of iron. That is why this alloy was recommended for production of castings for narrow walls of the mold.

3. Conclusions and results

On the basis of the results of the investigation, the following conclusions can be made:

1. The research group determined the liquidus and solidus temperatures of the processes of melting and crystallization of low Cu – Fe alloys, which depend on the iron content in these alloys.
2. All Cu – Fe alloys with the iron content of $\leq$ 0.3 wt. % have a finer grained microstructure.
3. It was found that all low Cu – Fe alloys with the iron content $\leq$ 0.3 wt. % have a 20% higher hardness, than Copper 1.
4. It was experimentally found that the temperature of primary recrystallization of low Cu – Fe alloys containing less than 0.3 wt. % of iron increased to T_{pk}= 360 °C.
5. The thermal conductivity and the temperature conductivity of low Cu – Fe alloys at temperatures of up to 400 °C is 1.5-2 times lower, than these of pure copper, while at the temperature of $\geq$ 600 °C, these characteristics are about the same.
6. The performed investigations proved that the low Cu – Fe alloys containing 0.17-0.3 wt. % of iron can be used for production of mold walls.

References

[1] Lyakishev N P and Shalimov A G 2002 *Development of technology for continuous casting of steel* (Moscow: OOO NP «Eliz»)

[2] Tischenko P I, Timoshenko S N and Kanfer V D 2001 *Proceedings of the 6 Congress of steelmakers* (Moscow: Chermetinformation) pp 288–291
[3] Ermolyuk T D, Lyakx A P and Tselikov A A 1985 *Steel* **6** 33–36
[4] Logunova O S 2008 *Steel in Translation* **38** 849–852
[5] Osintsev O E and Fedorov V N 2004 *Copper and copper alloys. Domestic and foreign grades: reference book* (Moscow: Machine building)
[6] Bogdanov S A 2004 *Metal supply and sale* **2** 51–57
[7] Motohisa M, Mosamitsu S, Hirosi N and Masaaki I 1988 *Japan «Metallurgy»* **2** 229–233
[8] Zaiman J 1962 *Electrons and phonons: Theory of transfer in solids* (Moscow: Publishing center of foreign literature)
[9] Dubskiy G A, Vdovin K N, Nefedyev A A and Dubskaya T Ya 2015 *Vestnik of Nosov Magnitogorsk State Technical University* **20** 81–88

IOP Conf. Series: Materials Science and Engineering **718** (2020) 012022 doi:10.1088/1757-899X/718/1/012022

On Capabilities of People with Health Limitations in Terms of Industrial Business

O A Denisova[1] **and O L Lekhanova**[1]

[1] Cherepovets State University, 5 Lunacharsky street, Cherepovets, 162600, Russia

E-mail: denisova@indox.ru

Abstract. The Resource Training and Methodical Center (RTMC) for training people with disabilities, based at the Cherepovets State University, presents its experience and research results in terms of identifying the capabilities of people with disabilities as a human resource for industrial business. The authors demonstrate that over the past few years the number of people with disabilities focused on inclusive employment in industry has been increasing. The specifics of careers chosen by people with disabilities and their special needs prove that cooperation between industrial business and a network of Resource Training and Methodical Center for training people with disabilities is a must.

1. Introduction

Over the past few years focus on inclusion of people with health limitations (HL) into every sphere of life of the society has become one of the key trends of governmental initiatives in Russia. Industrial business is no exception. Taking into consideration that ratification of Convention on the Rights of People with Disabilities evoked serious amending of legislative and regulatory framework in our country, issuing of decrees and taking of governmental solutions stimulating businessmen to study the capabilities of people with disabilities more thorough as a source for development of their factories and industries, analyses of the existing experience in terms of professional orientation, education, employment and social involvement of people with disabilities becomes quite relevant. This task becomes even more significant in relation to improvement of management framework of industrial manufacturing and its human resourcing, which is associated with the known controversy between the necessity to meet the governmental requirement for quota arrangement and filling of vacancies with the people with disabilities and the existing labor contraindication at the hazardous and dangerous industrial manufacturing of people with dysfunction of vital functions.

This research has been made as a part of activities of Resource Training and Methodical Center for training people with disabilities based at the Cherepovets State University established by the decree of Ministry of Education and Science of the RF (RTMC NWFD CSU). Such training is implemented at the Universities of the designated area of the North-Western Federal District (Arkhangelsk region, Vologda region, Kaliningrad region, Murmansk region, Novgorod region, Pskov region, Karelia republic, Komi republic) [1, 2]. RTMC NWFD CSU is a member of the Federal network of Resource Training and Methodical Centers (RTMC) for training people with disabilities based at the organizations of higher education (OHE) [3]. The conjoint efforts of RTMC network members is aimed at ensuring the equal rights for education, social and cultural integration and social involvement for the people with disabilities and HL, their inclusion into labor and social relations in the region [4].

IOP Conf. Series: Materials Science and Engineering **718** (2020) 012022 doi:10.1088/1757-899X/718/1/012022

This article presents a summary and systematizing of experience and practices of RTMC based at the Cherepovets State University related to identifying the capabilities of people with health limitations as a human resource for industrial business taking into account current data and opinions in the problem field of research [5-9].

2. Objective

To analyze the current status in the sphere of inclusive professional orientation, inclusive education and inclusive employment in the region and to define resources of people with health limitations for the purposes of industrial business.

3. Materials and methods of research

This research was made using methods of comparative analyses, systematizing and learning from the experience of RTMC NWFD CSU and evidence of analytical reports from the center activity for the period of 2017-18. The above mentioned objective was sourced by:

- regional statistics on the number of people with disabilities, including those people who are getting professional education;
- data on the number and proportion of students with disabilities in the Universities and organizations of higher education located at the designated area of North-Western federal district reserved for the RTMC NWFD CSU (Arkhangelsk region, Vologda region, Kaliningrad region, Murmansk region, Novgorod region, Pskov region, Karelia republic, Komi republic);
- data on varieties of health limitations among the students with disabilities in the partners' Universities of RTMC NWFD CSU;
- list of training programs chosen by the students with disabilities in the Universities of the designated area reserved for RTMC NWFD CSU;
- results of monitoring control among the students with disabilities at the designated area reserved for RTMC NWFD CSU (Arkhangelsk region, Vologda region, Kaliningrad region, Murmansk region, Novgorod region, Pskov region, Karelia republic, Komi republic);
- data on the estimated number of graduates with disabilities at the regional labor market.

4. Research results

The North Western federal district occupies 9,5 % of the territory of Russia, the proportion of population is more or less the same. According to the official report the total population of the region makes more than 13 mln people, 71,3 % of them belongs to the economically active people. Other 12 % is shared by the people with disabilities, of them approximately the third part is made by people of active working age, about 5 % of the total number of people with disabilities is shared by the young people aging from 18 to 30 years.

The statistics analyses of the data provided by FKU (Federal public institution) Chief Sociomedical Examination Board in Vologda region belonging to the Ministry of Labor and Social protection of Russia for the period 2017 – 2018 shows the following number of people with disabilities: in 2017 - 101642 people, in 2018 – 97419 people. In2017, 210 people with disabilities aging 18 – 30 years were getting professional education out of the total 3101, in 2018 – 261 people out of the total 2857.

Thus we can see the decrease of the total number of people with disabilities within the period 2017-2018 together with the increase in the number of those who are getting professional education constituting the human resource for industrial business. Special attention shall be paid to the data about people with disabilities who are getting higher education and are the most probable candidates to be granted the disability jobs quota in regard to the vacancies for the specialists of high qualification. See Table 1 for the data on the number of students with disabilities in the partners' Universities of RTMC NWFD CSU.

Thus the total number of students with disabilities who study in the partner's Universities located at the designated area of RTMC NWFD CSU is increasing. But apart from growth in the number of students with disabilities, their proportion in the total amount of the students studying at the

Universities is increasing. These data reasonably demonstrate both interest of the young people with disabilities in professional education and necessity of cooperation between industrial business and Universities practicing inclusive education and training the employee pool for inclusive employment.

Table1. Number of students at the partner's Universities of RTMC NWFD CSU in 2017-18.

Students with disabilities	2017		2018	
	Students with disabilities	Proportion in the total number of students	Students with disabilities	Proportion in the total number of students
Total in the network of RTMC NWFD CSU	331	0.27%	425	0.47 %

Data analyses on distribution of the students with disabilities of different years of study at the Universities of RTMC NWFD CSU network is as follows: 150 students with disabilities – 1styear; 2ndyear – 83, 3rdyear – 60, 4thyear – 57, 5thyear – 15, 6thyear – 3 students with disabilities. Consequently the further years shall see almost triple increase in the amount of graduates with disabilities at the regional labor market, which shall be considered by HR services when they address allocation of quotas and filling vacancies designated for the people with disabilities.

Data on nosology of dysfunctions among the students with disabilities in the partner's Universities of RTMC NWFD CSU see in Figure 1.

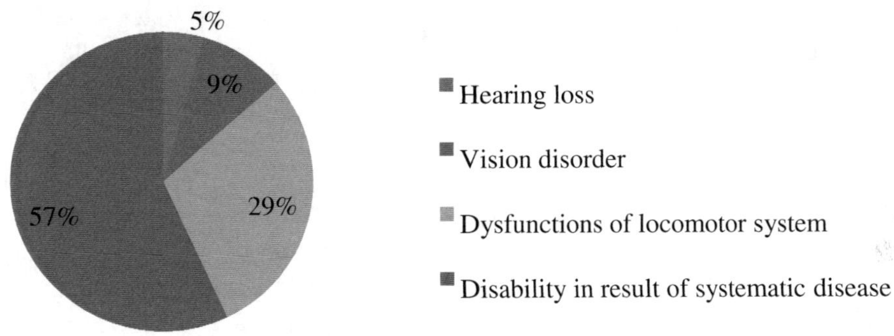

Figure 1. Nosology of dysfunctions among the students with disabilities.

As it becomes clear from the picture, most part of the students with disabilities suffer from systematic disease and dysfunctions of the locomotor system; the least number of students have hearing loss and eye disorder. It shall be noted that even inside of one nosological group the list of health limitations of the students with disabilities is non-uniform varying from minor need in special environment for studying and support to the need in continuous tutor's support and being trained according to the adapted educational program of general professional education. The opinion poll among the students with disabilities which was carried out in 15 universities of NWFD demonstrated that intramural studies together with students who have no dysfunctions is the most in-demand and convenient option (71%). 75% of students with disabilities point out the necessity to modify and adopt training programs so that to consider physical condition. Together with that the largest portion opinion poll participants identified demand in social and psychological support (45%), development of individual training program (18%) and assistance of a certain associate (18%). 79% of respondents mentioned that they don't need any specialized equipment, 2/3 of them instead pointed out the necessity to get social and psychological support. The available data are valuable as a prospective aspect to facilitate finalizing of the issue related to equipping of special work places and creating the special labor conditions for graduates with disabilities. When considering employment of the people with disabilities representatives of industrial business shall realize the needs of such applicants, including their need in mentoring and support, which shall be rendered by the people specially trained for cooperation with the people with disabilities and serving their special needs.

Analyses of monitoring data acquired from the students with disabilities at the designated area reserved for RTMC NWFD CSU (Arkhangelsk region, Vologda region, Kaliningrad region, Murmansk region, Novgorod region, Pskov region, Karelia republic, Komi republic) demonstrated that the largest portion of University applicants select the University at the maximum proximity to the place of their residence. Most of the people with disabilities address to opinion of their parents and teachers, fields of interest and hobby, data available at the official web-sites of the Universities as the ground for selection of educational program and major. Almost none of the respondents with disabilities defined their physical condition and recommendations provided by the Sociomedical Examination board as the basis for selection of the future profession. Rare is the judgment that the regional labor market and availability of vacancies to be employed for shall be accounted for while selecting the major. Monitoring of the educational needs of University applicants with disabilities demonstrated that only half of the high schoolers with disabilities (53 %) have clear understanding on the profession which they would like to be trained for. Third portion of respondents (32 %) have doubts on selecting the future careers. Alongside with that 46 % of the applicants say they don't need any assistance of employment counsellor. This means that University applicants with disabilities have just a vague idea on their future careers. In general, the acquired data reflect the regional tendency for disbalance between the socio-professional and value-based orientation of the population and structure of demand for the highly qualified jobs.

Data analyses on the selection of the future careers indicated that over half of the existing 98 training programs, being covered by the students with disabilities refer to professions intended for industrial business. The list of majors (specialities) mostly demanded among the students with disabilities at the Universities of the designated area reserved for RTMC NWFD in terms of their potential to be applied further in industrial business is shown in Figure 2.

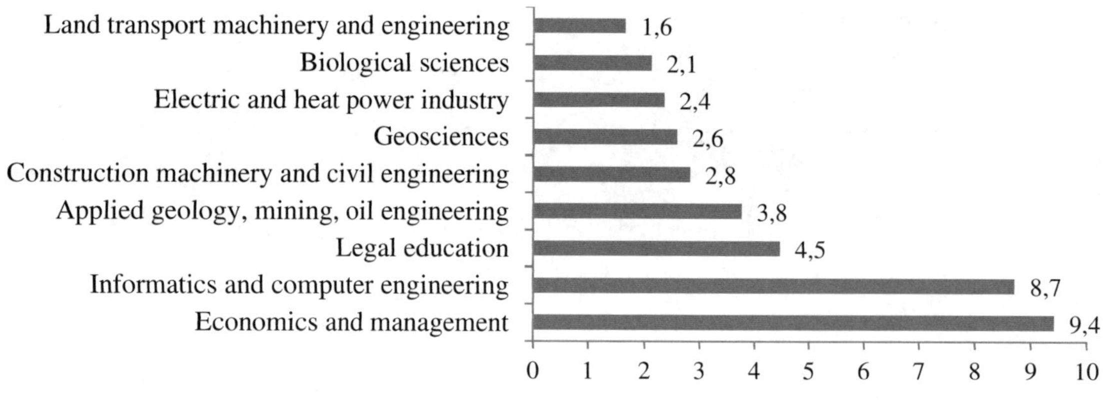

Figure 2. Top of majors (specialities) mostly demanded by students with disabilities focused at the industrial business in the partners' Universities of RTMC NWFD CSU (% of the total selected amount).

With that said the top of most demanded training programs (specialities) among the people with disabilities is the following: construction, information systems and technologies, applied informatics, economics, management, legal education, pedagogy education, psycho-pedagogy education, specialized (defectology) education, medical care). These training programs refer to the Extended groups of majors (training programs) "Engineering, technologies and technical sciences", "Social sciences", "Education and pedagogy sciences", "Health care and medical sciences" which shall be

IOP Conf. Series: Materials Science and Engineering **718** (2020) 012022 doi:10.1088/1757-899X/718/1/012022

taken into account when making a forecast for employment of people with disabilities at the regional labor market.

5. Conclusion and Summary

Practical experience of RTMC NWFD CSU and research results indicate that currently the portion of people with disabilities who are getting professional education with a focus at the inclusive employment is demonstrably increasing. Demanding mentoring and support at the workplace is notably varying among the graduates with disabilities and tend to have prominent individual character. As a whole there is a need for building an integrated system to coordinate activities of educational organizations, employers (including industrial business), social welfare institutions and labor authorities in terms of professional orientation, education, employment of people with disabilities and their adaptation at the workplace.

Addressing to the people with disabilities and HL as human resource for industrial business will be efficient under condition that a complex of interrelated data on inclusive professional orientation, inclusive education and inclusive employment is taken into account.

It appears that cooperation between industrial business and educational organizations which practice professional training of the students with disabilities and HL, also as a part of federal project of RTMC network activities will allow to:

- make real the employment of people with disabilities at the work positions for specialists of higher qualification limited by quotas;
- plan and reduce costs on equipping of specialized workspaces for people with disabilities and HL;
- improve the image of organizations and enterprises being socially focused;
- generate new competencies with their employees in terms of cooperation with people with disabilities and mentoring of their labor;
- increase the level of corporate culture including the issues related to generation of its inclusive aspect.

References

[1] Afanasiev D V, Denisova O A and Lekhanova O L 2018 *Developing inclusive higher education: the Network approach* (Moscow: MGPPU) pp 72–82.

[2] Denisova O A and Lekhanova O L 2017 *Psychological Science and Education* **1** 119–129

[3] Margolis A A, Rubtsov V V and Serebriannikova O A 2017 *Psychological Science and Education* **1** 10–17

[4] Rubtsov V V 2016 *Psychological Science and Education* **1** 87–93

[5] Bikbulatova A A, Soldatov A A and Nevskaya M V 2015 *Vestnik UGUES. Science, education, economics. Series: Economics* **3** 36-40

[6] Bryantseva M V 2011 *Social Policy and Sociology* **10** 82–88

[7] Natsun L N 2017 *Bulletin of UrFU. Series: Economics and Management* **4** 663–680

[8] Nenakhova Yu S 2018 Population **3** 96–108

[9] Shabunova A A and Fakhradova L N 2016 *Economic and social changes: facts, trends, forecast* **6** 126–142.

AUTHOR INDEX